TURNING SILICON INTO GOLD

THE STRATEGIES, FAILURES, AND EVOLUTION OF THE TECH INDUSTRY

Griffin Kao

Jessica Hong

Michael Perusse

Weizhen Sheng

Apress®

Turning Silicon into Gold: The Strategies, Failures, and Evolution of the Tech Industry

Griffin Kao
Wynnewood, PA, USA

Jessica Hong
Edina, MN, USA

Michael Perusse
Cambridge, MA, USA

Weizhen Sheng
Markham, ON, Canada

ISBN-13 (pbk): 978-1-4842-5628-2
https://doi.org/10.1007/978-1-4842-5629-9

ISBN-13 (electronic): 978-1-4842-5629-9

Managing Director, Apress Media LLC: Welmoed Spahr
Acquisitions Editor: Shiva Ramachandran
Development Editor: Rita Fernando
Coordinating Editor: Rita Fernando

Cover designed by eStudioCalamar

Distributed to the book trade worldwide by Springer Science+Business Media New York, 233 Spring Street, 6th Floor, New York, NY 10013. Phone 1-800-SPRINGER, fax (201) 348-4505, e-mail orders-ny@springer-sbm.com, or visit www.springeronline.com. Apress Media, LLC is a California LLC and the sole member (owner) is Springer Science + Business Media Finance Inc (SSBM Finance Inc). SSBM Finance Inc is a **Delaware** corporation.

For information on translations, please e-mail rights@apress.com, or visit http://www.apress.com/rights-permissions.

Apress titles may be purchased in bulk for academic, corporate, or promotional use. eBook versions and licenses are also available for most titles. For more information, reference our Print and eBook Bulk Sales web page at http://www.apress.com/bulk-sales.

Any source code or other supplementary material referenced by the author in this book is available to readers on GitHub via the book's product page, located at www.apress.com/9781484256282. For more detailed information, please visit http://www.apress.com/source-code.

Printed on acid-free paper

To our family, friends, and the summer of 2019.

Contents

About the Authors

Griffin Kao has a background in computer science, mechanical engineering, and economics from Brown University. He's worked as both a software engineer and product manager at B2B and B2C tech companies, both large and small, and has acquired a lot of insight into how the industry operates.

Jessica Hong studied at Cornell University where she received a bachelor's degree in computer science. She's been able to gain a deeper understanding of the impact of technology in the world through her previous roles as a software engineer, product manager, and venture capital partner.

Michael Perusse, Harvard University class of 2020, studied computer science and mathematical sciences. He's worked in product management at both Google and Microsoft and at start-ups, and has written for a variety of magazines and web sites.

Weizhen Sheng studied networked and social systems engineering for her bachelor's degree and data science for her master's degree at the University of Pennsylvania. She has previously worked as a product manager and software engineer in the tech industry and as an analyst at an investment board, offering her varying perspectives on the role of technology.

Introduction

So you're curious about how the tech industry operates? Then you've come to the right place! In this book, we present you with stories of both blinding success and utter failure, as well as of odd and unexpected phenomena, that paint a picture of a Silicon Valley simultaneously capricious and rational in nature. We've packaged these narratives in 25 easy-to-digest cases that each focus on a concept relevant, but not obvious, to anyone exploring a career at the intersection of technology and business.

In particular, we've grouped the cases into the five sections that comprise this book. The first, "Getting Started," largely analyzes how budding companies or individuals generate the initial spark. The second, "Gaining the Edge," focuses on the middle part of a company's storyline marked by growth and strategic perfecting. The third, "Extending the Lead," brings the discussion to mature companies, their dominance, and the problems they now face. Finally, the last two sections, "Failing" and "Society," respectively, consider what contemporary failure in the tech industry can look like and ways in which the tech industry interacts with broader societal dynamics.

In choosing the topics for these cases, we tried to avoid those we felt were already evident or self-explanatory. At the same time, we attempted to speculate as little as possible, providing analysis for past events and their implications, rather than predicting the future. For those coming from backgrounds in computer science or software engineering, you'll notice that we've abstracted away much of the implementation details when talking about technical solutions. We did this on purpose, since we believe that while technology itself certainly drives the narrative of the tech industry, it's the decisions made by humans regarding how technology is used or distributed that provide the most unique and transferable insights (and may be the most interesting to read about)!

We had a lot of fun writing this so we really hope you enjoy reading.

Getting Started

Part 1

Getting Started

The Second Mover Advantage

What Facebook, the iPhone, and Airbnb All Have in Common

Like that of the Roman Empire many years before, the meteoric rise of MySpace beginning in 2003 was sweeping and absolute. Although MySpace was preceded by several smaller social networking sites like Friendster and SixDegrees, none could match the user base, the press coverage, and the valuation given to MySpace by investors. For all intents and purposes, MySpace was the forefather of our modern-day notion of the "social media" platform—the behemoth with over 100 million monthly active users and millions, if not billions, of dollars in ad revenue.[1] In fact, MySpace was purchased after a huge bidding war by the renowned News Corporation (parent company of the Fox Broadcasting Company) for $580 million in 2005, one of the largest acquisitions of the time.[2] And in 2006, the social networking site surpassed Google as the most visited web site, then continued to dominate the social network space for the next couple of years.

---

[1] https://martechtoday.com/despite-ongoing-criticism-facebook-generates-16-6-billion-in-ad-revenue-during-q4-up-30-yoy-230261
[2] http://news.bbc.co.uk/2/hi/business/4695495.stm

© Griffin Kao, Jessica Hong, Michael Perusse and Weizhen Sheng 2020
G. Kao et al., *Turning Silicon into Gold*,
https://doi.org/10.1007/978-1-4842-5629-9_1

The Rise of Facebook

In theory, MySpace was destined for everlasting greatness—too big to possibly fail, just like the Roman Empire in its heyday. And yet, also like the Roman Empire, a competitor (the Byzantium Empire) laid in waiting to deliver a hard reality check. At the same time MySpace was enjoying its time in the limelight, Harvard students Mark Zuckerberg and Eduardo Saverin were quietly expanding their social network, initially called TheFacebook and limited only to Harvard students, to other colleges. In 2005, after having changed the name of the platform to just "Facebook," Zuckerberg's company received $12.7 million in funding from Accel Partners, and then in 2006, Facebook became available to the general public.[3]

Facebook's growth since opening to the general public was explosive, and in 2008, the site overtook MySpace in the Alexa rankings, which are based on Internet traffic. The next year, Facebook surpassed MySpace in unique monthly visitors to the web site. In the years after, Facebook continued to gain users, while MySpace saw its user base plummet—forcing a series of layoffs, management changes, and revenue losses. In 2011, News Corporation finally threw in the towel and sold the site for $35 million, a fraction of the price they originally paid.[4] Today, MySpace is largely relegated to the nostalgic memories of an era long past. The platform has redefined itself as an entertainment site that provides music, video, and news content with 15 million monthly visitors as of 2016, a far cry from the social media titan that it used to be.[5] At the same time, Facebook has grown to 2.41 billion monthly active users in the first quarter of 2019 and has a colossal market cap of $581 billion.[6] Despite being a latecomer to the social media space, Facebook is the clear winner over MySpace.

Breaking Down the Second Mover Advantage

By conventional logic, Facebook's triumph is counterintuitive. How many potential start-up founders have hesitated to enter a space because "someone has already done it"? However, this is actually a well-documented phenomenon called "the second mover advantage," the advantage a company obtains by following others into a market. There have been numerous examples of a market follower prevailing over a pioneer—Airbnb over CouchSurfing, the iPhone over the Blackberry, Gmail over Hotmail, Zappos over Shoebuy, and the list goes on and on. The second mover advantage has given these challengers the ability to learn from the mistakes of their

---

[3] www.cnet.com/news/facebooks-valuation-the-cheat-sheet/
[4] www.theguardian.com/technology/2011/jun/30/myspace-sold-35-million-news
[5] https://adage.com/article/media/myspace-juice-left-publishers/303781
[6] https://newsroom.fb.com/company-info/

predecessors, as well as ride the tailwind of a space already carved out in the "free-rider" effect. More specifically, these can be broken down into three distinct benefits:

1. **Product development and management:** Second movers save a lot of money on R&D (research and development), being able to work off of a product that has already been tested and optimized. Newcomers to an existing market can extrapolate the winning features and learn from failed launches or product gaps. Facebook, for example, modeled their platform after MySpace's, even choosing the same colors (blue border, white background). However, the company learned from MySpace's cluttered and unsightly wasteland of over-personalized pages, opting for a cleaner look for user profiles. And arguably more important, the platform listened to users where MySpace didn't, offering users a product to support what already mattered to them instead of trying to give them new things to care about (i.e., a forum to post and interact with other users rather than specified sections for things like karaoke or classified ads).

2. **Market acceptance:** Second movers don't have to expend resources on educating target users on the benefits of the product. Consumers are slow to change, so it can often be costly to convince them that a solution really works. The first mover has the responsibility of moving the market through the introduction stage of the product life cycle, while the second mover can go directly to growth and sales (Figure 1-1). When Facebook was first introduced, users were familiar with the idea of the social networking site—they knew how the product could help them and how to use it. In some cases, like Facebook's, market acceptance can even mean that investors are more willing to invest in the enterprise, accelerating growth and inflating value.

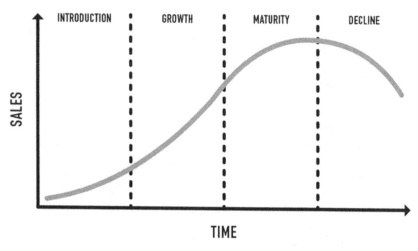

Figure 1-1. The Product Life Cycle

3. **Customer acquisition:** Second movers generally spend less on customer acquisition, because they don't have to spend the time and money testing multiple channels for sustainable acquisition. Instead, they can copy the channels already used to acquire new users like MySpace's feature for importing contacts or sending invites to the platform over email. Facebook took this concept and ran with it, not only allowing users to invite friends to join but also to do all kinds of engaged actions like liking pages or playing Farmville on the platform. Moreover, second movers can fine-tune the messaging of their predecessors. Facebook, for example, took MySpace's slogan, "A place for friends," and modified it to be more widely appealing. By claiming the platform "helps you connect and share with the people in your life," Facebook appealed not just to teenagers but to a wider-reaching audience.

Most importantly, the late mover gets to know that there *is* a market in the first place, without having to do any validation—the company first to market has already done all the heavy lifting. Some wildly successful companies, including Instagram, Kickstarter, Google, Nike, along with the aforementioned Airbnb, Apple, and Zappos, have enjoyed some or all of these benefits by following someone else to the market.

What This Means for You

At the same time, there are some obvious advantages to being the first to a market, like no initial competition, the potential for patent, and customer loyalty/brand recognition. In addition, not all markets are equally conducive to second movers. Some of the factors that make a market more friendly to latecomers include low customer investment (making it easier to switch between brands), long expected lifetime for the market, a rapid innovation speed allowing followers to iterate and improve on their predecessors, and product fluidity encouraging followers to borrow product features from a pioneer. Together, these characteristics make a market ripe for entry to second movers. A great example of markets like this is the shared electric scooter economy comprised of companies like Lime and Bird.

It makes sense that the first to a market is disproportionately valued by innovators and investors since there's high upside in the opportunity to gobble up market shares. However, there's also considerable risk involved in introducing your product to customers who might not find it valuable or who might be resistant to mass adoption. Entrepreneurs shouldn't give up on an idea just because there's a product out there that does the same thing or something similar. Embrace the second mover advantage to hedge risk and accelerate growth, because you could be the next Facebook, iPhone, or Airbnb.

Takeaways: Being the second company to enter or a general follower to an emerging market can oftentimes be cost-effective and less risky than being the first.

Questions

1. Are there times in your life as an individual when it's been helpful to copy or follow someone else? Are there times when it's been better to be the first to do something?

2. If you're looking to start a company, is the market you'd like to enter more conducive to first movers or second movers?

3. If you're looking to start a company, has there been someone else that has already attempted the same or a similar solution? If so, did it succeed and/or why did it? Did it fail and why did it? How can you use this knowledge to strengthen your plan and/or pivot?

Election Meddling

How One Man Changed an Election

In a vast world, we're often confronted with the reality that any individual action or set of actions we take is insignificant in the grand scheme of things. This idea affects decisions we make on a daily basis. At times, it hands us the courage to proceed without shame, like the many cases of *What difference does it make if I throw this wrapper on the ground?* At other times, it gives us pause and impresses upon us inaction. *Why does it matter that I drive an SUV gas-guzzler when millions of other Americans do? Why should I speak out for what is right when my voice is one in a billion?* Our inaction in the face of transgression is what enables the tragedy of the commons—the tragic depletion and destruction of a resource because individuals act independently, according to self-interest, and in contradiction to the common good. The feeling of insignificance nourishes evil and kills dreams; it causes environmental damage and widespread poverty. Indeed, this is the very basis of voter apathy. Contrary to all this, one man was able to expose the fallacy behind the thought by single-handedly influencing the outcome of the hotly contested 2017 special Senate election in Alabama. This is David Goldstein's story.

© Griffin Kao, Jessica Hong, Michael Perusse and Weizhen Sheng 2020
G. Kao et al., *Turning Silicon into Gold*,
https://doi.org/10.1007/978-1-4842-5629-9_2

Goldstein's Experiment

David Goldstein set out to shift the tide toward Democratic candidate Doug Jones who was running against Republican Roy Moore—a tall task given that the last Democratic win in a US Senate race in the state was in 1992. Inspired by Cambridge Analytica—the London-based company that leveraged questionable data mining/analysis tactics and targeted marketing to aid Donald Trump's 2016 presidential campaign[1]—Goldstein wanted to see if he could replicate their digital strategy in effectiveness with limited resources. In addition, he wanted to see if he could do so while remaining in compliance with legal and ethical guidelines. And to do that, he wanted to run a real-world experiment, as a proof of concept, with $85,000 in funding from individual donors. By comparison, the average cost of winning a Senate seat in 2016 was $10.6 million, according to the Center for Responsive Politics, a nonprofit, nonpartisan research group.[2] And that's just spending by the campaign itself—adding in external spending (outlays by party committees, etc.) nearly doubles that figure.

The crux of Goldstein's plan was a demand-side platform (DSP), software that allows for the management of advertising across many real-time bidding networks like Google Ads or Facebook Ads. DSPs typically allow marketers to bid in real time on ads, track posted ads, and optimize for cost and effectiveness, all in one interface. Goldstein paired the DSP with supply-side platforms (SSPs), network-specific tools like Facebook Connect, to enable programmatic advertising, the application of complex algorithms to automate the process of purchasing and posting successful ads to social media platforms and news sites at scale (see Figure 2-1).

[1] www.reuters.com/article/us-facebook-cambridge-analytica-factbox/factbox-who-is-cambridge-analytica-and-what-did-it-do-idUSKBN1GW07F
[2] www.opensecrets.org/news/2016/11/the-price-of-winning-just-got-higher-especially-in-the-senate/

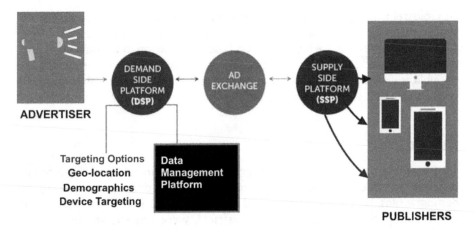

Figure 2-1. A DSP and SSP Pairing

In Goldstein's experiment, he split up the voter population into four groups targeted by separate campaigns (according to his own blog post explaining the experiment):[3]

1. **Conservative GOP voters:** These voters who clicked on the ads were taken to a site that showed them articles written by conservative or religious figures arguing against Roy Moore—since Goldstein believed that conservative and evangelical were underexposed to people with similar beliefs to them opposing Moore.

2. **Moderate GOP voters:** The moderate Republican voters were shown a different site that encouraged them to write in a different, more moderate Republican candidate (in opposition to the staunchly conservative Moore).

3. **Democrat voters:** Registered Democrats who clicked on the ads were taken to a site that asked them to commit to voting and where they could find where to vote. The site also displayed content published on other web sites about how Democrats were close to victory in Alabama depending on turnout.

4. **Unregistered but likely Democrat voters:** Finally, unregistered Democrats who clicked on ads were taken to register to vote online.

---

[3] https://medium.com/@david.goldstein_4168/https-medium-com-tovolabs-proof-of-digital-persuasion-in-alabama-senate-race-85a517481371

To Goldstein, a large part of the experiment was proving that he could achieve similar potency in his ad campaign to that of companies like Cambridge Analytica but without having to rely on false information in his published content—so he made sure that the content he was displaying to Democrats and Republicans alike was reliably true. Instead of altering the actual content, he wanted to make sure he was altering the flow of information to voters.

Impact

To determine the impact of the experiment, Goldstein selected three senate districts that could be matched against three other districts controlling for demographics, past voter history, and media consumption. Those in the control group, people in the first three districts, were not shown the ads, per targeting restriction that he could apply in the DSP. In contrast, he was able to get significant penetration in the test group, displaying a digital ad at least once to about 50% of the target demographics in those states. Although the click-through rate (CTR) was below average at 0.25% (the benchmark CTR for advocacy ads on Google Adwords is 1.6%)[4], Goldstein still observed 11,000 unique engagements on the curated experiment sites.[5] Breaking down the results even further, Goldstein saw even more direct evidence of his impact on the election. He witnessed a decrease in voter turnout among Republicans in the 3 experimental districts when compared to the control group: a 2.5% decrease for moderate GOP voters and a 4.5% decrease for conservative GOP voters. Conversely, he saw a 3.9% increase in voter turnout among registered Democrats in the same districts when compared to the control group.

Of course, verifying the accuracy of Goldstein's observations of his experiment's effect on the election is nearly impossible, but what's objectively true is that Doug Jones won the election, becoming the first Democrat in 25 years to win a US Senate seat in Alabama. While Goldstein's singular vote probably would not have changed the election results, Goldstein was arguably successful at changing the course of history by leveraging ads tech tools—all with a fraction of the resources typically consumed in a successful Senate campaign.

---

[4] www.wordstream.com/average-ctr
[5] https://medium.com/@david.goldstein_4168/https-medium-com-tovolabs-proof-of-digital-persuasion-in-alabama-senate-race-85a517481371

Technology's Role

We should emphasize the role technology played in this story, both augmenting the influence of Goldstein's actions and automating the execution of his agenda. Moreover, technology allowed him to run the experiment completely remotely, from New York City, where he was living at the time. His narrative is inspiring in the implication that our actions and words can be significant. In this new age dominated by viral content and information exchange across invisible networks, our voices can be amplified a million times over from any geographic location given the right Internet tools, allowing us to exercise more power than ever before in actuating change. However, this also means that technology allows individuals to cause harm at a greater scale—think data breaches or cyber attacks or deepfake revenge porn. So while Goldstein's story is an inspirational one, we must remember that the same strategy has been and can still be used to push nefarious agendas like ISIS propaganda or Russian interference into US elections.

Takeaways: Individuals should no longer frame the significance of our actions in the context of a tech-less world, since we can leverage technology for scalable impact. From a manager's standpoint, empower your employees with the tools to do more—but also with checks to prevent large-scale wrongdoing—and you'll see that small teams can provide considerable output.

Questions

1. What are some ways in which you currently use technology to amplify your voice or increase impact?

2. What are some situations in our individual lives where we might hesitate because our own impact seems inconsequential? And how can technology enhance our impact?

3. How can society, institutions, or individuals make sure ad technology is not used for "bad" purposes? What qualifies a purpose as "bad"—criminal, unethical, something else?

Cashless Cash

M-Pesa and the Rise of E-Currency in Africa

Dominated by vast swaths of savannah, lakelands, and mountain highlands, Kenya's landscape is beautiful. These ecosystems support a wide range of flora and fauna that create a backdrop juxtaposed against a much less prosperous economic climate plagued by issues arising from widespread poverty and political instability. Indeed, a profusion of child labor, the lack of access to education, and rampant prostitution are all largely related to the fact that 42% of the Kenyan population lives below the poverty line according to UNICEF.[1] And at a higher level, the financial institutions and tech industry of the African nation are consequently outmatched by those of the United States by nearly every relevant metric—from investment volume to rates of adoption for new technologies. It's no surprise then that we tend to think of "Silicon Valley" or Wall Street before Kenya as financial or innovation hubs.

Nonetheless, before Apple or Google or Square were able to push their respective forms of mobile payment to widespread adoption in the United States, people in Kenya were going cashless at shocking rates through the mobile payment system called M-Pesa. In fact, as early as 2013, 43% of the Kenyan GDP was already moving through M-Pesa, and over 237 million transactions were taking place on the platform.[2]

---

[1] www.unicef.org/kenya/overview_4616.html
[2] www.forbes.com/sites/danielrunde/2015/08/12/m-pesa-and-the-rise-of-the-global-mobile-money-market/#4b41014c5aec

© Griffin Kao, Jessica Hong, Michael Perusse and Weizhen Sheng 2020
G. Kao et al., *Turning Silicon into Gold*,
https://doi.org/10.1007/978-1-4842-5629-9_3

What Is M-Pesa?

Owned by the Nairobi-based mobile network operator Safaricom, M-Pesa is a mobile money transfer system that leverages the SIM card and its associated account with the service provider as a bank account containing some digitally maintained balance. Users can access their balance by sending/receiving money over the service, depositing/withdrawing cash, or checking their balance, all through SMS text messages and a preinstalled phone menu called the "SIM Toolkit" with select actions. To send money, a user would send a PIN-secured text message to the recipient denoting the amount to transfer, and both would receive a confirmation text message. Note that the recipient is only required to have a mobile phone and does not need to subscribe to Safaricom's service in order to receive the payment. While the user's updated M-Pesa balance is listed in the confirmation SMS for both sender and recipient, they can also request a text message with their balance at any time.

In addition, M-Pesa users deposit and withdraw cash through a network of "banking agents" consisting of brick-and-mortar retail outlets and airtime resellers (which resell call time that can be used on a mobile network). To put money into their account, a user visits an authorized M-Pesa agent who uses a specified "agent phone" to send e-money in exchange for cash or a transfer from the user's bank account. To take out money, users go through a similar process; the agent will give the user cash in exchange for an equivalent digital transfer from their M-Pesa account, making a note in a physical "logbook" that serves as a record of all transactions completed in the store. If a recipient to a mobile transfer isn't a registered user, they'll receive a one-time voucher with a 4-digit code with which they can exchange for cash at an M-Pesa agent.[3]

Of course, M-Pesa users aren't limited to sending money to other individuals and can actually pay vendors for goods or services through an additional feature called "Lipa Na M-Pesa" (Swahili for "Pay with M-Pesa"). Although there's a small fee associated with sending money and withdrawing cash, all other services provided by the platform are free to use. Once a user deposits money, the money theoretically doesn't need to be withdrawn until it's spent (at a low transactional cost), and in this way, M-Pesa serves as an e-currency allowing users to go completely cashless.

---

[3] www.safaricom.co.ke/personal/m-pesa/getting-started/using-m-pesa

The Rise of M-Pesa

It all began in 2002, when researchers funded by the UK's Department for International Development (DFID) observed that people in several African countries were transferring mobile airtime as a substitution to real currency. In particular, a common use case they noticed in Kenya was the transfer of airtime to relatives who would then use or resell the airtime. The researchers wanted to better facilitate this informal economy of airtime as a currency through a mobile-enabled system of authorized airtime credit swapping. Partnering with the British multinational telecommunications conglomerate Vodafone, they began piloting the service in 2005 in Kenya. Vodafone had already been brainstorming how to enable microfinance through mobile technology to address the arduous process of transporting cash as well as generally limited access to banking in the West African nation.

In 2007, the service was launched to the public by Vodafone's Kenyan subsidiary Safaricom and immediately began to change Kenya's economic landscape, especially for less wealthy Kenyans. The percentage of low-income Kenyans living on less than $1.25 a day who used M-Pesa skyrocketed from less than 20% in 2008 up to 72% by 2011.[4] More generally, M-Pesa created unprecedented financial inclusion by extending access to financial services to 20 million Kenyans.[5] With better access to capital and the financial world, such Kenyans were able to better save money and endure financial emergencies, fundraise for causes like disaster relief and education, and found new ventures powered by microfinancing. In fact, a study published by MIT economist Tavneet Suri, who has followed the financial and social impact of services like M-Pesa since 2008, has noted a considerable reduction in poverty—particularly among female-headed households. Suri's study estimates that access to mobile money services dilated daily per capita consumption for 2% of Kenyan households, pulling them above the $1.25 a day threshold that marks extreme poverty.[6]

M-Pesa's surge in usage was so substantial (see Figure 3-1) that formal banking institutions began to take note of the venture, considering it a serious competitor. Indeed, a group of banks unsuccessfully lobbied the Kenyan government in 2008 to audit M-Pesa in the hopes that such regulation would slow the platform's growth. In order to compete, traditional banks have turned to offering cheaper and more useful mobile banking services, many with transaction fees even lower than those offered by M-Pesa—making the market for mobile banking and cash transfer even more accessible.

---

[4] www.forbes.com/sites/danielrunde/2015/08/12/m-pesa-and-the-rise-of-the-global-mobile-money-market/#3065fb9e5aec
[5] www.forbes.com/sites/danielrunde/2015/08/12/m-pesa-and-the-rise-of-the-global-mobile-money-market/#3065fb9e5aec
[6] http://news.mit.edu/2016/mobile-money-kenyans-out-poverty-1208

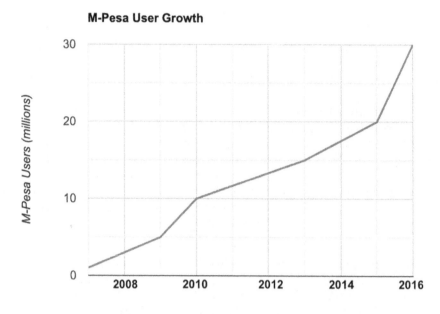

M-Pesa User Growth

Source: Vodafone, Safaricom

Figure 3-1. M-Pesa User Growth by Year

Empowered by the explosive growth in cell phone access in the developing nations, M-Pesa has also gained rapid traction in other countries—in Q1 of 2015, more than 900 million mobile users in Africa and 3.7 billion mobile users in Asia were registered on the service.[7] It's undeniable that M-Pesa was not only a remarkable success as evinced by sheer market penetration but also a crucial player in the adoption of mobile technology as a vehicle for economic transactions. Case in point—by 2014, there were about 255 mobile money platforms servicing 89 countries, according to the Global Mobile Systems Association (GMSA).[8]

Why M-Pesa Was Successful

While some of M-Pesa's market success can be attributed to favorable economic and regulatory conditions in Kenya, the role of their execution strategy—a strategy that focused on removing friction in the customer adoption process, on building trust, and on resolving the chicken-and-egg problem (see Chapter 5) through aggressive and simultaneous acquisition on two fronts—cannot be understated.

---

[7] www.forbes.com/sites/danielrunde/2015/08/12/m-pesa-and-the-rise-of-the-global-mobile-money-market/#3065fb9e5aec
[8] www.gsma.com/mobilefordevelopment/wp-content/uploads/2015/03/SOTIR_2014.pdf

The founders of M-Pesa recognized that barriers to using M-Pesa occurred at all stages of the customer conversion funnel (see Figure 3-2). First, they knew that in order to get Kenyans to consider trying the service, they had to capture them in the awareness stage of customer conversion. This meant not only reaching as many people as possible but also making it obvious in their messaging what M-Pesa did and why users should try it out. So Safaricom invested a lot upfront in diversified marketing channels and logistical infrastructure to power a nationwide launch of 750 stores, covering 69 district headquarters.[9] Their goal was to acquire 1 million customers within a single year, which was 17% of the 6 million people forming Safaricom's customer base at the time.[10] Since it was commonplace for younger family members to move to cities to generate income to send home while the rest of their family remained in rural areas, Safaricom's initial marketing efforts leveraged both TV and radio ads to target wealthier city-dwelling Kenyans with a narrative of young people sending money home to their families. This heartstrings-tugging campaign continued to reiterate the "send money home" idea that unified the M-Pesa brand to potential users and conveyed the use cases for the product. Their ads juxtaposed the inconveniences and perils of having to transport cash over long distances with the ease of sending digital money through M-Pesa's platform—and depicted the convenience of going cashless.

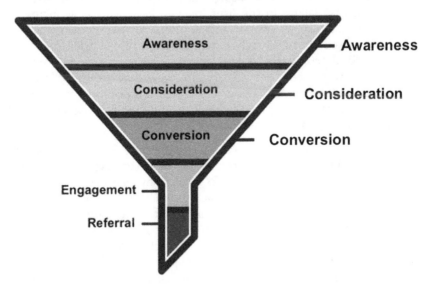

Figure 3-2. The Customer Conversion Funnel

[9] http://siteresources.worldbank.org/AFRICAEXT/Resources/258643-1271798012256/M-PESA_Kenya.pdf
[10] http://siteresources.worldbank.org/AFRICAEXT/Resources/258643-1271798012256/M-PESA_Kenya.pdf

Then to get users through the consideration and conversion stages, Safaricom made it nearly impossible not to try out M-Pesa. They made the service free to register and to deposit and opted to not specify a minimum deposit. To retain customers and build loyalty, they made the user interface incredibly simple with only a few actions accessible through the main menu of the user's phone, which could be the cheapest, lowest-tech phone offered. Moreover, Safaricom ingeniously killed two birds with one stone through the send-money-to-anyone feature. To registered users on the platform, this was great because it allowed them flexibility in transferring money, but Safaricom also charged them a higher fee to send to non-registered users. The hope was the sender would put pressure on the non-registered recipient to register for the service and the recipient, who just had a great first experience with M-Pesa, would be happy to oblige. The final stage of the customer conversion process, advocacy and referral, was thereby made trivial with the power of the network effect. And across all stages, the removal of adoption barriers allowed M-Pesa to reach the critical mass necessary for explosive growth.

Further yet, Safaricom recognized that trust was not only a critical friction in converting users but also necessary to create an enduring brand. To build trust, they ensured that their pricing structure was easily understood and known by all users—since a user being charged when they weren't expecting it was a worst-case failure in the user journey. Fees were only incurred at times where it made sense (when sending money or withdrawing cash) and were listed in fixed amounts rather than percentages. In addition, charges were standardized across the banking agents, which again could be physical retailers like convenience stores. Safaricom knew that if they depended on the outlets to collect charges, this could create the potential for abuse and variation, so they deducted fees directly from the user balance and provided commissions directly to the banking agents.

Then once a clear pricing structure was established, Safaricom worked to build user trust in the banking agents themselves so that users were comfortable handing over or taking out cash. To do so, they incorporated checks like a confirmation SMS text message for each transaction that served as a signal that the transaction was completed and a receipt to be used in the case of a dispute. The SMS would list the recipient and amount, allowing users to recognize if they've sent the wrong amount or transferred money to the wrong person. Safaricom also invested heavily in the standardization of the customer experience across multifarious banking agents, implementing training and supervision through a rating system. A third-party vendor would visit agents each month to evaluate them on a set of criteria related to the branding and experience of making M-Pesa transactions. In addition, each bank agent was required to maintain a record of all deposits and withdrawals in a standard Safaricom-branded paper logbook.

Perhaps most importantly, while M-Pesa was acquiring new users, the platform had to make sure it simultaneously nailed down the outlet side—getting stores to buy into a complex network of banking agents. To lay the groundwork for this, Safaricom conceived of a structure to not only manage the individual agents but also to incentivize them to contribute to the system. They employed intermediaries, called "Agent Head Offices" (agent HOs), that served as liaisons and performed the crucial functions of distributing commissions to the agents and making sure the agents had the cash to service M-Pesa customers. Safaricom paid agent HOs for each transaction conducted in stores under the agent HO's management and left it to the agent HO's discretion on how to split the commission with the agents. Of course, Safaricom was careful to balance an increase in the number of agents with growth in the M-Pesa user base since too many agents would reduce the number of transactions at each one and disincentivize being part of the network.

The Failures of Apple Pay

Let's briefly examine Apple Pay as representative of efforts by better-resourced tech companies in the United States to go cashless. To be clear, Apple Pay and its peers from other large tech companies, like Google Pay, Square, and Samsung Pay, are likely still on the path to widespread adoption. However, three years past its 2014 launch (August 2017), Apple Pay had reached just 12% of the iPhone user base.[11] It's undeniable that growth has been slow, especially when compared with the rapid penetration of M-Pesa. Apple Pay's slow growth is in part the result of vastly more developed economies and government infrastructure that made for far less favorable market conditions. Whereas M-Pesa successfully addressed the primary pain point of the taxing journey to transport cash, Apple Pay simply hoped to save users the need to carry around a credit card, a minor inconvenience.

Replicable Parts of the M-Pesa Strategy

If we were to distill the parts of the M-Pesa jumpstart strategy that are replicable, we would focus on how they minimized friction at all stages of the customer conversion process. The barriers to acquiring users and retaining them will look different depending on the market and the product, but the simple principle of always making it easier for a potential user to become a customer for your product or service can oftentimes be lost in the cacophony of other competing considerations like monetization or branding.

---

[11] www.statista.com/statistics/755625/iphones-in-use-in-us-china-and-rest-of-the-world/

We can also derive some insights from the ingenious management and incentivization structure Safaricom employed in constructing their committed network of banking agents. They made sure to utilize intermediaries, standardize protocol to give users the customer experience they expected, and manage growth.

Finally, perhaps the most important lesson M-Pesa can teach us is to understand your target market. An unreceptive market may ultimately be a barrier to adoption that cannot be resolved. So truly consider the market conditions in which you're launching your product or service—what kind of regulations can you expect? How strong is the user need you're addressing? What other competitors are out there? Because no matter how many resources you have at your disposal and no matter how effective your growth strategy, the course of the customer conversion process remains fundamentally in the hands of the market.

Takeaways: If you're starting a company with a product or service, make sure to remove all frictions in the customer conversion process, leverage the right incentive structure for users and other stakeholders, and truly understand the target market.

Questions

1. What are some other reasons why Apple Pay adoption was slow?

2. If you have a product or service, what barriers to adoption exist at the awareness stage? How can you overcome those? Likewise, which barriers occur at the consideration and conversion stages?

3. What are some effective strategies for leveraging the network effect once you've acquired a user? How can this differ across different industries?

Pivot to Perfection

How YouTube, Groupon, and Instagram Can Help You Pull Off a Successful Pivot

When YouTube was first founded in 2005, it was meant to be a dating platform. Though it is now a video-sharing platform with over two billion monthly active users,[1] it started off as a dating site with zero users at all. In fact, when it first launched with the slogan "Tune in, Hook up," the platform could not get a single person to upload a dating profile video onto the site. Even after the founders posted ads on Craigslist offering $20 to women who uploaded videos of themselves onto the site, no one would. So, in order to make it seem like YouTube had some semblance of a user base, one of the founders, Jawed Karim, resorted to posting a truly random assortment of videos, including footage of Boeing 747 airplanes and the now infamous video "Me at the zoo." After these uploads, the platform began to take on a new form. People began to use the site not for dating but to share and watch tons of different videos: ones of their pets, vacations, seemingly anything. In response, the YouTube team worked to completely revamp the web site to facilitate much more general content.

---

[1] www.youtube.com/about/press/

© Griffin Kao, Jessica Hong, Michael Perusse and Weizhen Sheng 2020
G. Kao et al., *Turning Silicon into Gold*,
https://doi.org/10.1007/978-1-4842-5629-9_4

A mere five months later, YouTube fully pivoted to the general video-sharing platform we know and love today.

Much like YouTube, many successful companies have needed to course-correct their business strategies. Pivoting happens to various degrees in a variety of ways, and in the following sections, we'll take a deeper dive into when companies decide to pivot and how they do it successfully.

Knowing When to Pivot

To understand how companies are able to pivot successfully, we need to understand why they pivot in the first place. During the course of any company, you always have three options: to persevere, to pivot, or to pull the plug. The biggest signals to pay attention to when deciding what to do are your customers' actions and your long-term viability. It's important to continually gauge these things, even if your company seems to be doing well on the surface.

First off, paying attention to the actions of your customers is crucial to understanding if a change in course is necessary. Paying attention to user behavior can help you understand how people actually use your product, telling you which use cases to double down on and which to drop. You can track this behavior with both qualitative user research and quantitative metrics. Though qualitative user research is useful for getting a nuanced view of user behavior, quantitative metrics are important because customers may sometimes voice opinions that are inconsistent with their actions. Even if survey says your customers like your product, actions speak louder than words, so you need to see cold, hard evidence in the form of sales and adoption to know if you are on the right path. Thus, when the customer acts, you should pay attention.

Finally, even if customers are happy and short-term goals are being met, you need to look out for long-term business health. If anything you are doing in the short term is hurting you in the long term, then the path you are on is not sustainable. While every company, particularly early-stage start-ups, has experienced the need to fight for survival—to get through the next month, week, or even day—you need to ensure that your short-term efforts are not at the expense of long-term viability. If your near-term efforts are creating a mess you need to clean up later on, it may be time to pivot.

In the case of YouTube, it was largely the customer behavior that indicated the need for a pivot. No one was signing onto the platform for the dating profiles; instead, users were completely ignoring the intended use case and uploading their own random assortment of videos. By listening to their customers, YouTube followed a strong signal indicating it was time for a change, leading to an ultimately much more successful product.

User-Driven Pivots

YouTube demonstrates one particular type of user-driven pivot a company can make, though there are many different ways a company can pivot as well. User-driven pivots occur due to the behavior of different user groups and fall into three subcategories: the product pivot, customer pivot, and problem pivot.

Product Pivot

The product pivot occurs when a company changes their product focus because user research shows more opportunity with a different set of features. This necessary user insight can be collected through many different channels—from tracking metrics to direct user observation. For example, YouTube noticed the change in use case simply by browsing the videos being uploaded on their platform. A different web site may monitor user behavior by tracking the number of visits to different pages, shifting their focus based on which pages receive the most engagement in the form of views and clicks. On the whole, such user insight clarifies the product market fit of your current product offering and can generally push you toward a better product.

Product pivots often come about when a specific feature dominates total user engagement. For example, Instagram experienced this method of product pivot, originating as the app Burbn. Burbn was a location-based app that allowed you to check in with friends, post pictures, and make future plans. It was well-received but had issues with widespread adoption because of its overwhelming jumble of features. To fix this, Burbn decided to listen to its users. Among all the different features available, the majority of people who used the app loved it for its photo posting and sharing capabilities. The founders decided to double down on this use case and create a new product called Instagram, which gained immediate success, acquiring more users within hours than Burbn had in a year.[2]

Another way a product pivot can manifest is when users begin to use your product in new unintentional ways, such as with YouTube. Though the platform was meant for video dating, users ignored this and co-opted it for their own purposes, thus sparking the need for YouTube to shift its product and revamp the platform for general video-sharing purposes.

Additionally, product pivots can arise when a company makes small bets on product changes and one takes off. In fact, this is exactly what happened with Groupon, a web site that offers limited time deals from businesses to consumers.[3] Groupon was born out of a product called The Point, a web site that

[2] www.washingtonpost.com/news/innovations/wp/2015/07/02/the-7-greatest-pivots-in-tech-history/
[3] www.groupon.com/merchant/article/the-history-of-groupon

used social media to encourage people to donate money to certain causes or goals. The Point web site had been experiencing modest usage in Chicago but had little to no growth elsewhere. In order to gain more customers, one of the founders, Eric Lefkofsky, brought up the idea of using the platform to facilitate group buying to earn discounts on products. Once this was implemented, this feature took off, and people began to use the group buy feature much more frequently than the donation feature, pushing The Point to pivot to what is known as Groupon today. This product pivot was encouraged by user behavior, with customers' actions clearly voting on what features they liked and what features they didn't. This little bet allowed them to invest minimal effort on a product change that brought a lot of success further down.

In short, product pivots occur when users respond more positively to features than to your intended product. This type of pivot is highly driven by user feedback and can also fundamentally change the problem you are solving or the customers you are targeting. It is absolutely critical that you have the right methods for tracking metrics and gathering user insight. Without conducting user research and tracking metrics, a product pivot cannot succeed. As shown in the preceding example, product pivots can range from minor changes to drastic shifts and often happen during the early stages of a company while they are still working to find product market fit.

To see if a product pivot makes sense, answer the following questions:

1. Is one of your product's features stickier than others? Why?

2. How are people using your product the most?

Customer Pivot

The customer pivot occurs when a company decides to focus on a new segment of customers that values your product more than your current customers, for example, Eloqua launched as a messaging application for the financial services industry, where it was deemed functional but not essential to core business needs. Though this product worked well, it began to truly catch on after Eloqua pivoted to a different user segment: companies looking to generate new leads (new potential customers). Here, Eloqua found a great product market fit and was acquired by Oracle for $871 million dollars in 2012.[4] Through diligent user research, Eloqua managed to find a customer segment that valued its product more highly, allowing it to pivot to a much more successful product. In other situations, a company can unexpectedly see their product picked up by an unintended segment—so it is crucial to characterize customer segments and recognize product market fit within different segments as they develop.

---

[4] www.oracle.com/corporate/pressrelease/oracle-buys-eloqua-122012.html

The following question is helpful when assessing a customer pivot:

1. What other customer segments could gain value from your product? What is their customer lifetime value (CLV)?

Problem Pivot

The problem pivot occurs when companies decide to focus on a different problem area. Starbucks experienced a shift in problem focus after the company originally started as a retailer of coffee beans and coffee equipment. Inspired by the popularity of espresso bars in Milan, early employee Howard Schultz decided Starbucks should create a similar environment in the United States, shifting from not only supplying coffee but also producing coffee.[5] The first Starbucks cafe opened in Seattle in 1984, and customers responded extremely positively. This hugely successful experiment paved the way for the new company it is today.

When considering a problem pivot, think about the following questions:

1. Is the problem you are solving validated by user research?

2. What other problems do your users face?

3. Is your problem a core problem for your user?

Pivots for Long-Term Viability

Other company pivots are driven by the disconnect between short-term action and long-term business health. These sorts of pivots include pivots in technology, pivots in growth strategy, or monetization pivots. In general, pivots for longer-term viability tend to be less drastic than customer-driven pivots because changes to core product, customer segment, or problem space are not needed. Instead, better alignment of resourcing and overall strategy is needed to ensure that the company continues to grow as needed in the long term.

A technology pivot occurs when a company changes the way they build their solution, perhaps by changing the tools and software used to build their product. This may be done to achieve better long-term viability by using technology that is cheaper, more efficient, more scalable, etc.

A growth pivot occurs when a company decides to focus on a new strategy for growth, including new approaches to customer acquisition, retention, or engagement. This is often needed to sustain growth because growth channels

---

[5] www.starbucks.com/about-us/company-information/starbucks-company-timeline

will eventually become saturated, become expensive, or disappear over time. This can also describe a change in the overall growth engine. Three of the primary growth engines include viral, paid, and sticky growth models. Viral growth occurs when users spread information about a product through their network. Paid growth is using money to acquire and retain users in the case of buying ads or other forms of paid promotion. Sticky growth is when retention is high and churn rate (the percentage of users that stop using your product) is low. A shift in the primary growth engine that a company uses is also a growth pivot that can help align a company with their longer-term business goals.

A monetization pivot describes changes in how you monetize your product and make money. For example, YouTube made a monetization pivot when they decided to introduce paid subscriptions, where users can pay for premium content or features that are otherwise inaccessible, in addition to YouTube's primary monetization strategy through ads. These kinds of pivots can end up impacting aspects of your product, business model, sales, marketing operations, and more. This pivot also has a direct effect on business viability, as it deals with revenue and profit generation to sustain the company. As budding companies are increasingly starting off with free products, the monetization pivot is becoming a regular part of the company narrative.

How to Pivot Successfully

Overall, pivots are a natural part of a company's growth and evolution. In order to keep up with changes in the market or user behavior, companies must learn to adopt—nothing lasts forever. In order to pivot successfully, it is important to remember three key ideas:

1. **No idea is precious**

    You should never get too attached to your original ideas. If any of the companies listed earlier had stuck with their first ideas, you likely wouldn't be reading about them right now.

2. **Deliberately learn from your past**

    In order to pivot intelligently, you must have a good idea of what you are currently doing well and what you are doing poorly. Without a good understanding of what has gone wrong before, how can you learn from your mistakes and ensure they do not happen again in the future?

3. **Make little bets**

    Before diving into a change, try taking baby steps and
    experiment in order to validate a new business hypothesis.
    In order to pivot successfully, you want to ensure that
    your new idea is validated by data and user research, as
    well as the results of any experiments you can conduct.

Overall, many of the companies we know today have gone through, and will
continue to go through, pivots as they continue to grow. Pivots are a natural
part of a company's evolution and ability to stay relevant in an ever changing
marketplace. Putting thought and intentionality into the company direction is
the best way to ensure a successful pivot.

Takeaways: Companies such as YouTube, Groupon, and Instagram originated as completely different products. By paying attention to customer
responses, they were able to find better product market fit and pivot into the
companies they are today. Though there are many different ways to pivot a
company, the success of a pivot is determined by constant learnings and validation from users and data.

Questions

1. We explore many reasons a company may want to pivot.
    When might a company want to stay on its set course?

2. Many pivots are motivated by customer response to a
    product. Are there any times that customer reactions
    present a red herring or a bad-quality signal?

The Chicken or the Egg

How OpenTable and Amazon Use Single-Player Mode to Solve a Classic Marketplace Problem

When OpenTable was founded in 1998, it promised to revolutionize the excruciating process of making restaurant reservations. Founder Chuck Templeton first had the idea for this product after his wife spent "three and a half hours calling restaurant after restaurant"[1] trying to make reservations for an upcoming visit from her parents. After this grueling experience, he decided enough was enough, thus creating OpenTable. Thanks to OpenTable, you can book a table at some of the hottest restaurants in the country with the click of a button. It makes life much easier for the restaurants as well—they no longer have to manually juggle tables in an archaic reservation book, as OpenTable does it for them. OpenTable successfully created a marketplace to connect millions of hungry customers with thousands of restaurants. In 2014, the company was acquired by the Priceline Group (now Booking Holdings) for a whopping $2.6 billion,[2] and as of now, the platform boasts over 123 million

---

[1] www.americanexpress.com/en-us/business/trends-and-insights/articles/opentable-founder-chuck-templeton-on-starting-up/
[2] www.wsj.com/articles/priceline-to-buy-opentable-for-2-6-billion-1402660209

© Griffin Kao, Jessica Hong, Michael Perusse and Weizhen Sheng 2020
G. Kao et al., *Turning Silicon into Gold*,
https://doi.org/10.1007/978-1-4842-5629-9_5

monthly users.[3] To date, users have spent over $69 billion at the restaurants on its platform.[4]

Clearly, both diners and restaurants saw immense value in the product, but the online marketplace space is notoriously difficult. In particular, marketplaces are difficult because they aim to connect buyers and sellers, yet without buyers, sellers will not join, and without sellers, buyers will not join. As you can see from their current success, OpenTable was able to overcome this issue—but how?

Marketplace Problems: The Chicken and the Egg

First off, let's identify the general concepts at play. A multisided platform is a model in which two or more separate parties come together to transact on some network. It has become increasingly popular because of the increasing connectivity brought on by the rise of technology. For example, Google's ad business is a multisided platform where consumers use its product (Google Search) for free, advertisers pay for them to see their ads (Google AdWords), content publishers get paid to display ads (Google AdSense), and many other parties exchange data or impressions.

OpenTable is specifically a two-sided platform and one of the first platforms to fully leverage an online presence to facilitate an ecosystem. The two-sided platform is also known as a marketplace product because it connects suppliers to buyers. In this case, OpenTable connects restaurants and hungry customers by giving them a platform to interact with each other. OpenTable then charges a fee from the restaurants for bringing them customers and enabling the transaction. As a marketplace, they create value for the buyers by finding suppliers and create value for suppliers by finding buyers. See Figure 5-1.

Figure 5-1. How Marketplaces Produce Value

---

[3] www.bookingholdings.com/brands/opentable/
[4] www.opentable.com/about/

Herein lies the root problem of marketplace products—the chicken-and-the-egg problem. Without buyers, suppliers see no point in joining a platform where they would sell to nobody. Without suppliers, buyers have nothing to buy—one will not exist without the other. So which comes first—the buyers or suppliers? The chicken or the egg?

The Origin of OpenTable

Now that we better understand the problem, let's examine how OpenTable managed to solve it. When OpenTable first began, they didn't manage a single restaurant reservation. In fact, they didn't even work on the customer side at all. They started working solely with restaurants, focusing their efforts on creating a table management and customer relationship management product. At the time, most restaurants were using paper notebooks and recording things by hand. This stand-alone product known as the Electronic Reservation Book (ERB) provided software to digitize the manual processes of table management and successfully brought restaurants onto their platform. Once restaurants started using the ERB, they were hooked. OpenTable was able to monetize this software using a subscription model where restaurants paid a monthly rate for access to the ERB. After creating this already valuable product for restaurants, they saw even more potential in becoming a marketplace and reaching diners as well. With the ERB, they now had highly valuable information about restaurant seating inventory which they could display to customers, allowing them to make reservations. This was highly beneficial for restaurants as well, as OpenTable helped streamline their reservation process by leveraging data already uploaded by the restaurants. Now, OpenTable could make even more money by collecting a finder's fee from restaurants, transforming them into the product we know and love today.

OpenTable's Approach: Single-Player Mode

One popular strategy used by marketplace companies in approaching the chicken-and-the-egg problem is to target a single set of users first. This is exactly what OpenTable did. In particular, they saw an opportunity to initially target the supplier side. By creating the Electronic Reservation Book, they were able to provide value to restaurants regardless of the marketplace product. Once they were able to get enough restaurants on board with their platform—cha-ching!—they now had a compelling reason for consumers to join the platform: they were able to make reservations at a variety of popular restaurants.

This "attract-one-side" strategy, also known popularly as playing "single-player mode,"[5] has been used by a variety of now-successful businesses. In general, if a marketplace company is able to provide a "single-player mode" for a product (a way to use the product that provides value even without other buyers/suppliers), such as OpenTable's ERB, they have a way to gain the initial traction necessary to begin "multiplayer mode" with both sides of the market.

Other successful companies such as Amazon were also able to use this strategy to get the necessary user acquisition to begin a marketplace. Though we now know Amazon as one of the world's largest online e-commerce marketplaces connecting both buyers and sellers, it had its humble beginnings as an online bookstore. Their strategy was to attract buyers before sellers. Amazon would buy books from other retailers and resell to buyers on its platform. After attracting enough buyers onto the platform, a variety of different sellers were motivated to join the Amazon platform so they could list their products on the platform and set the prices themselves. After that, the rest is history.

Why Single-Player Mode Works

Both OpenTable and Amazon were able to create a "single-player mode" for their products to attract one side of the marketplace and generate early adoption for their products. But why does "single-player mode" work in the first place, and how can this solution be applied to other marketplace platforms?

First, this approach solves the chicken-and-the-egg problem by creating a simpler one-sided product space, letting you decide on a side to target first. Without the complications introduced by a marketplace, you can focus on creating a high-quality product to provide value for a particular group of users. Other problems such as user acquisition and retention are simplified because you now only need to worry about half of the picture. Though creating a successful product for even one side of the marketplace is by no means an easy task, it is definitely simpler than simultaneously creating a product for two sides of the marketplace.

Secondly, creating a "single-player mode" creates an initial way to monetize your platform, providing more resources to grow a marketplace further down the line. By creating cash flow early on, before even providing a marketplace,

---

[5] https://cdixon.org/2010/06/12/designing-products-for-single-and-multiplayer-modes/

OpenTable was able to generate subscription fees that allowed them to put even more money into acquiring supply and generating demand, further helping grow their marketplace. In fact, even 10 years after its founding, 54% of OpenTable's revenue still came from subscription fees, with 42% coming from the reservation system.[6] Creating "single-player mode" was not only a successful restaurant acquisition strategy but a large part of OpenTable's continued success and profit today.

Finally, companies that are able to create value through "single-player mode" may have lower churn (the percentage of customers that stop using your product within a certain time frame). Products that are initially used in "single-player mode" create extra friction to leave "multiplayer mode" because leaving the marketplace means leaving the entire product. This is especially the case when the tools in "single-player mode" make up a crucial component of business operations. In the case of OpenTable, once restaurants moved from paper reservation management to the Electronic Reservation Book, it became hard to go back and thus hard to leave the OpenTable platform.

Even with all these benefits, there is one caveat to the single player approach. As amazing as it might seem, there are simply many cases of the chicken-and-the-egg problem where this strategy cannot be applied. For example, YouTube is a marketplace for content creators and content consumers, but there is very little value in a single player mode for a video streaming platform. In these scenarios, different approaches to the chicken-and-the-egg problem must be used.

Overall, there are many reasons why this approach is particularly successful. By simplifying the marketplace problem, creating an early monetization strategy, and lowering churn, "single-player mode" effectively addresses the chicken-and-the-egg problem and provides other benefits to help create a successful marketplace.

Takeaways: The chicken-and-the-egg problem occurs when a platform needs two sides to be useful, but one side will not find value without the other. Using "single-player mode" to initially attract one side through a different value proposition is an effective way to solve the chicken-and-the-egg problem and gain the necessary early traction to create a successful marketplace.

Questions

1. What are the "single-player modes" of some other popular marketplaces, such as Airbnb or Lyft? Do they even have one? How many marketplaces have a "single-player mode"?

---

[6] https://techcrunch.com/2009/01/30/opentable-files-for-ipo-and-reveals-its-finances/

2. If "single-player mode" cannot be applied, what are some other approaches that can be used to solve the chicken-and-the-egg problem?

3. For companies with both "single" and "multiplayer" modes, which mode ends up being more critical to monetization and business success?

Gaining the Edge

The Secondary Market

Don't Wait to Cash Out on Your Venture

So you're a cunning and resourceful start-up founder, a younger Jeff Bezos, who has grown your company from two people working out of your parents' basement to a venture capital-backed major disruptor of X industry. Life is moving smoothly, except for one thing—it'll be *years* before you can IPO (performing an initial public offering, the event in which a company begins selling its shares on the stock market). You're not sure what company would be willing to acquire your fledgling company, but you want cash *now*. You want to see the fruit of your hard labor. What do you do? Turn to the secondary market!

You might be asking what exactly is the secondary market, a fair question, but before we begin to answer that, let's first take a step back to examine the industry of venture capital at a high level. The textbook definition tells us that venture capital is a type of private equity or funding provided by firms to promising new companies—but this robotic description fails to capture the truly exquisite machine that powers innovation and technological progress on a massive scale. Indeed, venture capital lays the groundwork for a robust pipeline of ideas to reality, empowering entrepreneurs with the freedom to dream big and produce solutions even bigger. In its simplest form, when someone has

© Griffin Kao, Jessica Hong, Michael Perusse and Weizhen Sheng 2020
G. Kao et al., *Turning Silicon into Gold*,
https://doi.org/10.1007/978-1-4842-5629-9_6

an idea for a new venture that needs money to start or to grow, they can pitch the idea to a venture capital firm that then provides them with funding if the firm believes the idea will one day pay back the investment plus some extra return.

Generally, the funding firm will ask for some stake in the venture as a percentage in exchange for the money. And then the venture will "exit" when it grows large enough and valuable enough to either IPO (see Chapter 16 for more details) or to get acquired by another company. In both scenarios, whoever has a stake in the venture essentially receives some payment commensurate to the size of their stake. Admittedly, venture capital is imperfect given that many impactful ventures lack the earning potential to pique the interest of firms driven by monetary return on investment—not to mention that economic incentive and the long-term health of a venture are often misaligned, enacting divides between founder and investor. But at the end of the day, much of the tech industry wouldn't be here without the lifeblood that is venture capital.

What Are Secondary Markets?

Primary markets broadly refer to markets in which assets are directly issued. These assets could be anything ranging from securities, like bonds or stocks, to physical goods, like cars. For example, consider the sale of tickets to an Eagles game on the Ticketmaster web site. These tickets are listed there by the franchise itself, so when a fan buys tickets, the proceeds from the sale will go directly to the Eagles front office—we say these tickets are bought in the primary market. Conversely, secondary markets refer to the subsequent exchange of commodities between parties independent of the initial issuer. In the same example, the resale of the Eagles tickets by an individual fan on a platform like StubHub contributes to the secondary market.

The venture capital primary market is the initial deal in which the company directly trades a stake or equity in the company to some venture capital firm in exchange for capital. So then the secondary market consists of various transactions between investors involving the already-issued equity in a private venture (see Figure 6-1). More specifically, some shareholder (who can be the founder, an early employee, or early investors) can get some buyer to pay them some amount equal to the perceived value of the exchanged shares. In order for shareholders to find secondary buyers, the company usually has already proven some form of value, likely through market traction (a large share of the market) or revenue (high sales volume or value). Of course, it's easy to see that such transactions are great for someone who invested in a venture early on, yet doesn't want to wait what could be many years to get their money back when the company goes public or gets acquired.

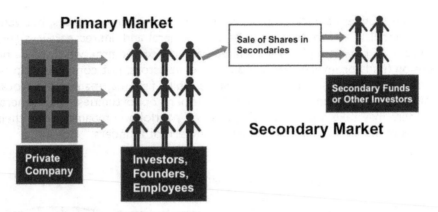

Figure 6-1. Venture Capital Primary and Secondary Markets

As a quick side note, it's important to point out that the terms "primary market" and "secondary market" can be confusing since they are often used to refer to when a company performs an IPO in the primary market and then continues trading their shares in the secondary market on stock exchanges like NYSE (New York Stock Exchange) or NASDAQ (again, see Chapter 16 for more details). However, the trading of such public shares on the stock market is distinct from the exchange of private shares in the venture capital primary and secondary markets—which are the ones we discuss here. In the private shares case, such transactions are subject to limited liquidity/cash exchange and often intense regulation by the SEC (US Securities and Exchange Commission) on who can buy the shares and at what quantity.

A Closer Look at Secondaries

To understand what happens in secondary transactions, we must first understand the types of shareholders that may be transacting and, in particular, the difference between a general and limited partner. A "partnership" is formed when multiple parties come together to manage or own a venture—this entity has no filing fees nor registration requirements, but it does have important implications for how liability is held. Within this partnership, each party may be either a general or a limited partner. A general partner can be seen as the owner of the partnership, actively managing the venture's day-to-day operations. In exchange for this power, they hold unlimited financial liability in the case of some adverse event like a lawsuit. A limited partner, in contrast, is removed from the daily proceedings of the company but maintains some stake in the venture. In general, limited partners are only liable for the amount of capital they invest in the business but are privy to some proportional percentage of the partnership's profits. While limited partners forfeit decision-making power, they get the benefit of sharing in the venture's growth with limited liability.

Start-up founders and/or employees are typically general partners, but venture capitalists and investors can be both general and limited partners (see Figure 6-2). For example, while limited partners to the firm, which could be pension funds or investment banks or angel investors, just contribute capital to the fund, a general partner at a venture capital firm may manage the allocation of the firm's funds by pursuing investment opportunities. Such general partners may take a close advising role for portfolio companies using their extensive industry experience to guide tenderfoot founders.

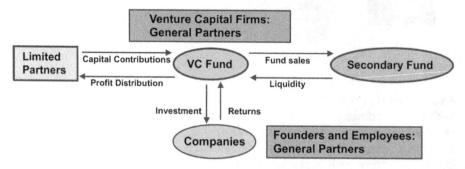

Figure 6-2. A Typical Partnership Structuring (Highly Simplified)

Now if we examine the mechanics of secondary transactions in light of these partnership structures, we see five common types of transactions:

1. **Direct secondary:** Direct secondary purchases are the most straightforward secondary transaction. The holder of the shares in a private company, such as a founder or employee or investor, sells their directly held equity to some buyer. To reiterate, it's a direct ownership stake being exchanged, rather than some fund interest (shares in a fund which in turn hold shares in portfolio companies) or other indirect ownership stake. Usually the equity being sold is highly illiquid, and the seller wants to make some form of profit in advance of the exit event of the company.

2. **Fund restructuring:** Put simply, in a fund restructuring, a general partner typically moves assets or shares from funds exceeding their expected life cycles to new vehicles (i.e., a new fund with different terms). Limited partners participating in the older fund can choose to cash out and sell their stake or transfer their shares to the new special-purpose vehicle. These vehicles, also called continuation funds, can provide general partners acting as portfolio

managers greater control over the portfolio and the ability to maximize the potentially underperforming remaining assets. Fund restructuring, along with direct secondary deals, were the most common type in 2017.[1]

3. **Stapled secondary:** Stapled secondary transactions are usually just fund restructuring or tender offer deals where the general partner will cleverly add another requirement for the purchasing parties to additionally commit fresh capital to a new fund. The general partner essentially "staples" on the obligation to make primary capital contributions.

4. **Tender offer:** In a tender offer, a general partner holds an auction (called "tendering") for currently held interests to incur offers from secondary buyers. Of course, limited partners holding those interests can then choose to liquidate or maintain their stake in the fund. Tender offers are typically public and set some price and time period during which the limited partners are invited to sell their stake. This offer price is typically higher than the expected or going rate.

5. **Preferred equity:** Often used in real estate investments, preferred equity deals usually have a general partner providing a debt-like interest as preferred equity to buyers. This allows general partners to maintain control and ownership of the underlying equity in the company, yet allows them to provide liquidity to limited partners. Additionally, preferred equity provides more flexibility than standard debt (i.e., borrowing from a bank) since there's minimal liability.

Such secondary transactions can provide the liquidity for general and limited partners to cash out on their book gains, as well as other benefits: reducing leverage (debt) on a fund, capping losses on underperforming assets, better allocating resources to stronger investment areas and eliminating nonstrategic investments, remaining up to date with the regulatory environment, and reducing liability from additional capital requirements. Regarding the last benefit, additional capital requirements or follow-on investments (investments made by an investor who made a previous investment in the company) are often required as a result of "pay-to-play" provisions in the term sheets governing venture capital deals.

---

[1] www.preqin.com/insights/blogs/gp-restructuring-a-growing-secondaries-market/24678

As for the buyers—while we can observe a variety of secondary investors, secondary funds that specialize in the execution of such transactions have grown increasingly common in the 21st century. These funds have realized that they can reduce risk by avoiding earlier-stage companies without a proven track record and instead buying shares in companies already on track to a traditional exit. While secondary funds typically make less per portfolio company than early investors, they're more than compensated with the reduction in risk.

Three major types of secondary funds include limited partner or LP funds, direct secondary funds, and hybrid secondary funds. The distinction between these is largely predicated upon the types of secondaries they prefer—LP funds primarily acquire indirect partnership interests through venture funds, direct funds perform direct secondary transactions per the name, and hybrid funds do both.

Growth of Secondary Funds

In recent years, deals in the secondary market have become larger and more frequent. For instance, in 2018, we saw one of the largest secondary transactions ever in which SoftBank paid close to $6.5 billion for an approximately 15% stake in Uber as part of a tender offer.[2] Additional large 2018 secondary deals include a $600 million secondary investment also in Uber; a $500 million purchase of shares of Credit Karma, the personal finance company; and another $500 million deal with the online legal tech company LegalZoom—each of which greatly exceeded in value the largest secondary transaction in 2017, a $250 million investment in Klarna, a Swedish company providing online financial services.[3]

So why have secondaries become more appealing? In a one-line answer, it's that stakeholders are eager to cash out. As we've mentioned several times, stakeholders don't want to wait for one of the traditional exits, IPOs, or M&A (mergers and acquisitions). But a closer inspection yields a few key drivers, very much in line with the secondary transaction benefits we discussed for general and limited partners, that underscore this impatience:

1. We've observed longer holding periods for private companies because of (1) a plethora of private funding opportunities in a friendly lending environment with low interest rates, (2) current macroeconomic conditions

---

[2] www.wsj.com/articles/softbank-succeeds-in-tender-offer-for-large-stake-in-uber-1514483283
[3] https://pitchbook.com/news/articles/alternative-exits-the-rise-of-secondary-deals-in-venture-capital

that make the public market less welcoming (a sequence of recently high-profile IPO flops; see Chapter 16), and (3) situational obstacles in finding the appropriate buyers for a venture. Here, we refer to the holding period as the time from initial investment to traditional exit. Indeed, the median holding period has risen to 8.2 years in 2017 from 3.1 years in 2000, while the term of early-stage investments is likely closer to 15–16 years.[4] It's important to note that general partners who have the long-term interest of the venture at heart, and thus see value in the continued private incubation of the company, may be at odds with limited partners who simply seek immediate payout. But at times, even for general partners on the venture capital fund side, longer holding periods may be incongruent with their investment strategy that both require relatively quick big wins on investments and synergy between revolving portfolio companies.

2. Coinciding with longer holding periods is the commonplace use of equity in lieu of cash to compensate the management team and early employees. In a world of high-profile tech billionaires, whose wealth can be traced directly to successful exits, stakeholders in the venture are happy to take the shares in the hopes of a greater payout down the road, but meager salaries then feed the impatience to see real cash in hand.

3. The flexibility derived from secondary transactions is valued more so now than ever—flexibility in capping losses to retain salvage value, and as we've reiterated several times, in deriving liquidity from growth of the venture made on paper. Just like an investor trading on the stock market who sees their stock in a company take a nosedive, general and limited partners may want to sell their holdings before they take too great of a hit, particularly when they don't believe it's possible to turn around the fortune of the venture. Not to mention, early investors, especially individuals and angel investors, can often take on losses from the dilution of their interest in later investing rounds. Without getting into the nuance of dilution, we can see that high-paying investors can demand larger stakes in the venture, pushing down the original holdings of previous investors.

---

[4] www.allenlatta.com/allens-blog/vc-time-to-exit-reaches-82-years-pitchbook

4. Private companies, or rather the management teams, increasingly see the need for operational liquidity to do things like pay employees, lobby for friendlier regulation, deal with lawsuits, etc.—and view secondary transactions as the means to obtain that liquidity.

5. Finally, with the deepening focus of early and maturing ventures on growth, especially growth in market shares or revenue, we've witnessed a shift toward multiple expansion strategies. In such strategies, an investor gets in on the venture and can flip their stake for some multiple of their initial investment (and a tidy profit) in the absence of true appreciation in value because some form of superficial growth in the company warrants a greater perceived value. In fact, an increase in user volume may correspond to no increase in profit or even net losses, but investors may not even care given that it only matters whether they have some other buyer to later off-load the investment.

More generally, secondary markets require some threshold of demand to sustain continued activity, demand growing alongside a growing venture capital industry. More specifically, we saw the industry grow in global value from \$53 billion in 2008 to an estimated \$160 billion in 2017.[5] And so, growth in the amount of private capital available has truly upped demand for shares in promising start-ups and by extension interest in the secondary market.

The Effects of an Expanding Secondary Market

For the economy at large, growth in the venture capital secondary market may not be a good thing, since this could be a sign of bubble-like conditions. To reiterate, if we revisit our last driver of secondary market activity, we notice that multiple expansion strategies are evocative of the oft-referenced 2008 housing crisis or even the "original" bubble, the 17th century tulip mania that plagued Europe. We've already seen ill omens in the form of companies with huge valuations due to massive growth truly struggling to generate a profit or conceive of a strategy to do so (see Chapter 16 for more details).

However, at the individual level, this is great news for stakeholders in a budding venture. Growing demand in the secondary market gives you more flexibility to liquidate your equity even with weaker evidence of underlying value. And when we take a step back, this is truly a beautiful phenomenon, since it

---

[5] www.telegraph.co.uk/technology/2018/11/20/global-venture-capital-industry-has-trebled-size-160-billion/

represents the impeccable fine-tuning of our great venture capital machine—the system that not only gives entrepreneurs the resources to devise solutions to our world's problems but also incentivizes them to do so with the tantalizing prospect of a large payout at the end of the road. In truth, the bottom line is that money talks and economic incentive is the greatest catalyst for innovation.

Takeaways: The secondary market, where stakes in venture-backed private companies are exchanged between different investors, represents a third exit option for stakeholders to cash out. Transactions in the secondary market are typically one of direct secondaries, fund restructuring, stapled secondaries, tender offers, or preferred equity deals. Secondary market activity is on the rise due to longer holding periods from initial investment to exit, among other factors. This may not be good for the economy as a whole, since much of this activity may be due to multiple expansion strategies which create bubble-like conditions but create powerful incentives for individual stakeholders at growing ventures.

Questions

1. What are other examples of primary and secondary markets in the world around you?

2. Which partnership structures and circumstances are conducive to each type of common secondary deal?

3. Do you think the venture capital secondary market should be more heavily regulated?

Cutting Out the Middleman

How Data Is Driving Netflix Originals

You're cuddled up on your bed, nice and toasty, having just finished the third season of *The Office* for the second time, when the inevitable "Continue watching" box appears. You sigh, knowing that as the countdown bar slowly dwindles, so does your will to do anything productive for the rest of the night. Season 4 is just as good as you remembered.

But as you and all of the other 139 million[1] Netflix users spend on average an hour each (1.15 hours, to be precise) watching their content each day, Netflix is watching *you*. No, your webcam isn't turned on, but all of the actions you take on the site are heavily monitored in hopes of painting a better picture of who *you* are as a viewer who's willing to stick around for the $12 subscription, for now. This relationship between consumers and technology companies has in recent years become unsurprising, but the extent to which Netflix drives the future of its entire platform's content development off of this data is remarkable from both a technology and business lens.

---

[1] www.cnn.com/2019/01/17/media/netflix-earnings-q4/index.html

G. Kao et al., *Turning Silicon into Gold*,
https://doi.org/10.1007/978-1-4842-5629-9_7

Before Netflix Originals

In 2013, Netflix tweeted that over 75% of its users made content selections based off of recommendations—the bread and butter of the technology entertainment powerhouse.[2] This includes any selection a user makes that doesn't involve them searching for a title on their own. In this regard, every Netflix account is unique, as no one's homepage will show the exact same selection of movies or TV shows. What's fascinating, however, is the level of detail that Netflix puts into optimizing your selection and consumption experience. As we'll continue to discuss, a hunger for better recommendations actually made Netflix originals the next logical step. In fact, that year, 2013, Netflix found itself at a stable enough point where it began to experiment with not only buying the (expensive) rights to stream content but also producing it themselves. Any content that Netflix produces itself—by funding the production and working with creatives—is a "Netflix original" series or movie. Because Netflix owns the content from its conception through production to airing stages, it does not pay any royalties for hosting these originals. *House of Cards* was the firstborn of only seven Netflix originals made in 2013. In just 5 years, Netflix would grow its original content family to have over 700 Netflix original TV shows and 80 original films on its site.[3]

Why Original Content

What contributed to such a boom in content creation? The answer, for any reasonable business, is that it was the right economic move.

From a business point of view, by making their own original content, Netflix could save money by moving away from owning expensive streaming rights. In addition, such existing content might not always be on the table, as other media giants begin to pull their shows and movies for their own, new streaming services, as is the case for both *Friends* and *The Office*. Both of these shows account for approximately 12% of all of Netflix's traffic, and will both be off of the site by 2021.[4] Netflix has been preparing for this potentially terrifying attrition of its user base by diverting resources to original content. In fact, over 20% of all of Netflix's streaming traffic lies in content that will be taken off the platform and moved to new streaming platforms from the likes of Disney, Fox, WarnerMedia, and NBCUniversal. Thus, Netflix needs to fill the impending void with other hit series that keep subscribers, well, subscribed.

---

[2] https://twitter.com/netflix/status/365577591563882496?lang=en
[3] https://variety.com/2018/digital/news/netflix-700-original-series-2018-1202711940/
[4] www.vox.com/2018/12/21/18139817/netflix-most-popular-shows-friends-office-greys-anatomy-parks-recreation-streaming-tv

Tech-Powered and Promoted Content

But beyond the fact that original content is cheaper to make and provides ownership, it also holds potent value to viewers as the company can leverage massive user data to craft a perfectly tailored product. From a technological point of view, Netflix originals can be incredibly *targeted*. Netflix has some 27,000 tags to label a given piece of content on their site, in addition to information such as how long a viewer watched a series, what series get binged the most, and what content is watched by a certain demographic of their subscribers.[5] Furthermore, Netflix can use its own users to help classify content; if X show gets many views from a demographic that likes sci-fi movies, then X show can gain that tag.

Additionally, Netflix is inferring things that go beyond your limited mouse clicks on their site; they're predicting things like race, age, and sexual orientation, all based on the content you interact with (or *don't* interact with). Using these data in aggregate across all users, they can statistically identify the kinds of content that keep a given demographic engaged with the site. And with the advent of Netflix originals, they can continue to feed and expand what their users demand.

For instance, if the data reveal that increases in content with the "Young Adult Romance" (a made up example but likely not far from reality) show an uptick in user retention of women in the United States and Canada aged 14–18 years of age, then that revelation trickles down to what scripts are green-lighted and what appears on the site.

The idea that in-demand content will be produced is not novel, but the extent to which Netflix can glean information from its corpus of users goes far beyond traditional TV ratings, which largely only consider view count (which itself is just an estimate).

Additionally, Netflix originals offer a unique opportunity to customize how the content is displayed on the site. The "poster" for a show is the thumbnail image of a show or movie that sometimes plays a video clip, trailer-esque summary of the premise when the user hovers over it or clicks into it. Given that Netflix is the creator and host of its original content, it affords them an opportunity to change how this poster is revealed to each of its users, dependent on their user profile.

To illustrate, imagine that Netflix has an original rom-com. To a user who Netflix believes to be a fan of comedies, the poster of the film might show a wacky image of the main protagonists and the trailer might highlight some of the best laugh lines and cute-but-awkward moments. On the other hand, to a

---

[5] www.finder.com/netflix/genre-list

user who Netflix believes to be a fan of romance movies, the poster and trailer are likely to be more romantic and sweet.

While explaining the technology, Netflix used the non-Netflix original example of *Good Will Hunting*, where comedy lovers were shown a photo of Robin Williams and romance lovers were shown a photo of the main lovers kissing. Imagine how much more freedom in creating appealing posters the company can have when they get to write the script, have their own photoshoots, and employ their design teams?

Problems with Targeted Content

While this choice of content advertising benefits Netflix by appealing to the different sensibilities of its users, there has been backlash as to how far such targeting has gone. Most notably, there have been several instances of Netflix original films with predominantly white actors/actress displaying posters exclusively showcasing the single black actor/actress to black viewers. Such targeting in this case is misleading and harmful to users who have responded with frustration to Netflix's targeted posters. This example poses many questions regarding how far a tech company should profile its users while still adding value to the suggestions and optimizations they're making.

Continue Watching?

Overall, Netflix originals have allowed for the company to throw lots of darts for cheap and see what sticks. Netflix has canceled dozens of originals after only one or two seasons yet has also found success in several series and films such as *Bird Box*, which had a record-breaking 26 million viewers in its first week on the platform. In 2019, Netflix spent an estimated $15 billion to produce its original content, and the number is only expected to grow amidst new streaming competitors entering the market, such as Disney+.[6] Netflix's ability to produce and host content has without a doubt changed the landscape of the entertainment industry. Whether or not its technological edge and original content strategy can save it from an increasingly competitive streaming market remains to be seen. Be sure to tune in for it.

Takeaways: Netflix uses user data to determine what content to create and how to promote it. Producing original content can be much cheaper than hosting popular existing content from third parties. Netflix has been shifting its focus toward original content due to its ability to effectively engage and retain demographics of users for the right cost.

---

[6] https://variety.com/2019/digital/news/netflix-content-spending-2019-15-billion-1203112090/

Questions

1. In what cases should a tech company have recommendation systems in place? What instanced could they implement recommendation systems but probably shouldn't?

2. What business risks are there in only creating content that targets certain niches?

Innovating an Industry

How Nintendo Single-Handedly Saved the Video Game Industry

You speed forward, deftly navigating around evil mushrooms and turtles that are coming to attack you. With a jump, you land on one of these mushrooms and manage to defeat it. You're currently traveling through the Mushroom Kingdom to rescue the Princess from an evil turtle, and the journey is taking you through several lands. Along the way, you find coins to collect and, if you're lucky, rare items like a Super Mushroom that double your size.

As your quest progresses, you notice a green pipe ahead. Curious, you climb into and slide through it. Upon emerging on the other end, you're shocked yet delighted to find yourself in a completely different world—you had just teleported via a warp pipe. You continue forward in this new world, knowing that more adventure lies before you on this journey to rescue the Princess. Exploration is endless.

This, of course, is the very familiar and beloved gameplay of *Super Mario Bros.*, envisioned and brought to life by Shigeru Miyamoto, Nintendo's legendary game designer. His iconic creation, the short, plump Italian plumber whom we all know and love, revolutionized gaming and saved the video game industry from a near-death experience. In short, *Super Mario Bros.* is both one of the greatest and most important video games in history.

© Griffin Kao, Jessica Hong, Michael Perusse and Weizhen Sheng 2020
G. Kao et al., *Turning Silicon into Gold*,
https://doi.org/10.1007/978-1-4842-5629-9_8

History of Video Games

Today, the gaming industry is thriving with new technological advances such as cloud computing and virtual reality applications. It can be difficult to imagine that this very same industry was close to collapse just a few decades ago during the 1980s. It was a classic case of market saturation, where too many companies were releasing too many poor-quality video games too quickly, causing consumers to lose faith in the industry. Simultaneously, personal computers were rapidly growing in popularity as an alternative to gaming consoles.

The result was an industry that shrank nearly 97% within the span of a few years,[1] dropping from an industry-wide revenue of $3 billion in 1982 to a measly $100 million in 1985 (see Figure 8-1).[2] The notorious 1983 video game crash can be attributed to multiple factors, including Atari's failed releases of *Pac-Man* and *E.T.* (with the latter often dubbed the worst game in history).[3] At a time that seemed like game over for the video game industry, Nintendo arrived and saved the day, much like Mario does in many of his games.

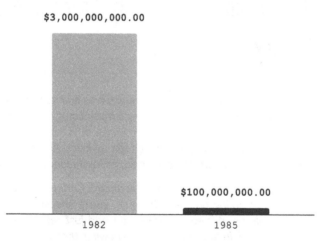

Figure 8-1. The Video Game Industry-Wide Revenues of 1982 vs. 1985

In 1985, Nintendo's launch of the Nintendo Entertainment System (NES) in the United States was met with applause and acclaim.[4] The console came bundled with *Super Mario Bros.*, which quickly became the "must-play" game of the NES and one of the best-selling video games in history, with over 40 million copies

---

[1] www.bugsplat.com/articles/video-games/great-video-game-crash-1983
[2] Steve Jones, Encyclopedia of New Media (SAGE Publications, 2002).
[3] Ibid.
[4] www.bugsplat.com/articles/video-games/great-video-game-crash-1983

sold (holding the record until Nintendo's later *Wii Sports* franchise overtook the title).[5] The console and game's resounding success revived the gaming industry and paved the way for the booming industry that exists today.

Shigeru Miyamoto: Nintendo's Legendary Video Game Designer

One can think of numerous reasons why Nintendo was able to successfully land the NES in the United States. They were smart with their marketing strategy and bundled *Super Mario Bros.* with the NES. They positioned the NES as an entire entertainment system, far more complex than just a console.[6] They introduced a "Nintendo Seal of Quality" to promise consumers good-quality games.[7] But perhaps most importantly, they were led by Shigeru Miyamoto's vision and willingness to innovate.

Most game designers in the 1970s and 1980s began the game development process by focusing on the technical, programming, and hardware aspects.[8] Shigeru Miyamoto approached it differently. He wanted to bring alive his childhood nostalgia, paying homage to his memories of wandering and exploring the fields in his backyard as a child.[9] Miyamoto arrived at Nintendo and brought along with him a new focus on exploration and storytelling in games. At a time where most video games were shooter or maze games, this was unheard of. His games—from the 1981 *Jumpman* (Mario's first incarnation, where he was involved in a bizarre love triangle with his pet, Donkey Kong, and princess Pauline) to his other immensely popular franchise, *The Legend of Zelda*—all tap upon people's inner urge for exploration and love for stories.[10]

Nintendo and the Art of Innovation

When one thinks of the word "innovation," one imagines unheard-of technological breakthroughs and advances. But innovation sometimes just means bringing a new paradigm to the world, taking something that's existed for years and finding a fresh application for it. Miyamoto understood humans' love for adventure and storytelling. He created the *experience* of video gaming— the experience of feeling immersed in a world that we can explore, with

---

[5] www.gamecubicle.com/features-mario-units_sold_sales.htm
[6] https://mashable.com/2015/09/13/super-mario-30th-anniversary/
[7] Ibid.
[8] https://schoolofgamedesign.com/project/shigeru-miyamoto-video-game-design/
[9] www.newyorker.com/magazine/2010/12/20/master-of-play
[10] https://schoolofgamedesign.com/project/shigeru-miyamoto-video-game-design/

smooth graphics, delightful music, and characters we can feel invested in.[11] He didn't introduce any groundbreaking technology but, rather, took the age-old concept of storytelling and innovated the gaming industry with it.

Innovation to delight users continues to remain in Nintendo's DNA today. Through the years, much of the gaming industry has consistently been pushed forward by baby steps in computation power; Microsoft's Xbox and Sony's PlayStation consoles constantly battle each other to claim the title of the world's "most powerful console."[12] Nintendo has always strayed from such competition, instead having its own distinct reputation as a company that pushes the boundaries of gaming. Nintendo has never failed to think outside the box and challenge its competitors to do so as well. Instead of purely relying on computation power, Nintendo has shown again and again the willingness to evolve and innovate the experience of gaming for users.

Beyond Miyamoto's creation of Mario, Nintendo has also contributed portable gaming (the 2004 Nintendo DS), motion gaming (the 2006 Nintendo Wii), and hybrid gaming (the 2017 Nintendo Switch) to the gaming world. None of these were necessarily using cutting-edge technology at the time; the Wii used accelerometer and optical sensor technology to creatively incorporate motion sensing into gaming,[13] more than compensating for the fact that its hardware specs lagged behind its competitors, Microsoft's Xbox 360 and Sony's PlayStation 3.[14] Instead of focusing on polishing its hardware, Nintendo directed its creative energy into revolutionizing gaming and making the Wii accessible for nontraditional gamers. This focus on noncustomers is often praised and is an illustrative example of the *Blue Ocean Strategy* (a marketing theory delineating the creation of a new market space and new demand).[15]

Even today, Nintendo is always able to stand out in the industry by demonstrating a willingness to approach gaming unconventionally and bring new innovative ideas to the industry. While technological advances and impressive hardware specs are certainly helpful, they can't substitute for true product innovation and creativity. Miyamoto once said in an interview that he always wondered where manholes led.[16] That curiosity sparked the birth of Mario and his pipe-teleporting adventures and introduced a sense of adventure that revolutionized the video game industry. Regardless of the domain or sector,

---

[11] www.techtimes.com/articles/217799/20180101/why-legendary-video-game-developer-shigeru-miyamoto-hires-non-gamers-to-make-games.htm
[12] www.telegraph.co.uk/technology/2017/11/03/xbox-one-x-review-does-worlds-powerful-console-live-name/
[13] https://hexus.net/tech/tech-explained/peripherals/19458-motion-sensing-game-controllers-explained/?page=2
[14] www.nytimes.com/2006/11/24/arts/24wii.html
[15] www.blueoceanstrategy.com/teaching-materials/nintendo-wii/
[16] www.newyorker.com/magazine/2010/12/20/master-of-play

creative thinking and product innovation are invaluable to companies aspiring to innovate an industry.

Takeaways: Companies can find success in straying away from what their competitors are doing (i.e., incremental hardware improvements) and instead focusing on innovating their products and bringing fresh ideas to the industry.

Questions

1. The 2014 Nintendo Wii U is often seen as a huge failure and can be described as the midpoint between the Nintendo Wii and the Nintendo Switch. Nintendo took risks with this console and attempted to tackle hybrid gaming, only to be met with a lukewarm market. What are some risks that come with innovation, and how can one recover from innovation failures?

2. Today, Microsoft and Sony are both multinational conglomerate tech companies, covering a variety of verticals, industries, etc. How does Nintendo, which is purely a video game company, compete and stay relevant in the industry?

3. Are innovations using groundbreaking technologies (i.e., cloud gaming) more impactful or innovations that creatively use existing technologies/concepts (i.e., the Nintendo Wii's use of existing motion-sensing technologies)?

Flourishing Freemium

How Slack Managed to Create One of the Highest Paid Conversion Rates in the Industry

When Slack came onto the scene in 2014, it promised to kill email once and for all. This collaborative instant communication platform was envisioned with the extremely lofty goal of completely revolutionizing workplace culture. Yet, while ambitious, it is hard to say that this goal has not been met. Used by 77% of Fortune 100 companies,[1] Slack is the fastest-growing enterprise software ever created, reaching 10 million daily active users and a 13.2 billion dollar valuation[2] in only 5 years. Users also just love the product, as is documented by the many thousands of notes of appreciation on Slack's Wall of Love[3] and the 3 million users[4] who put their money where their mouth is by actively paying for the service. In fact, this 30% conversion rate soars high

---

[1] https://usefyi.com/slack-history/
[2] https://news.crunchbase.com/news/working-to-understand-slacks-recent-valuation-declines/
[3] https://twitter.com/slackhq/timelines/402603838554644480?lang=en
[4] https://techcrunch.com/2018/05/22/slack-introduces-actions-to-make-it-easier-to-create-and-finish-tasks-without-leaving/

© Griffin Kao, Jessica Hong, Michael Perusse and Weizhen Sheng 2020
G. Kao et al., *Turning Silicon into Gold*,
https://doi.org/10.1007/978-1-4842-5629-9_9

above the industry standard, with Dropbox having a "really good" conversion rate of 4% and most normal conversion rates sitting around 1%.[5] With these mind-boggling statistics, you have to wonder, how did Slack manage to take on the Goliath that is email and become so successful?

From Glitch to Great

To understand Slack, you must first understand its CEO and cofounder, Stewart Butterfield, who grew up in a small Canadian fishing village named Lund,[6] which had a population of less than 300 people.[7] Yet even with this tiny population, or maybe because of it, Butterfield grew up fascinated by social interaction, in particular, social interaction through games because of their ability to bring people together. As he explains to Reid Hoffman on Masters of Scale, "it's not games that are so interesting to me, it's play as an excuse to interact with people socially."[8] This focus on community and social interaction is what pushed Butterfield to work on a series of games including Game Neverending, a massively multiplayer online (MMO) game that eventually failed and pivoted to Flickr.

After this first failing MMO, Butterfield was still determined to make a successful game. This next game, named Glitch, eventually gave birth to Slack. Though Glitch managed to attract a small and loyal fan base, it closed after interest in flash-based gaming plummeted. As Glitch began to bleed out cash, the company that owned the game (headed by Butterfield) decided to pivot and see what they could do with the rest of the money they had. Fortunately, they had a solid option for a new product. The Glitch team had developed an internal communication system to use while building the game, and they decided to focus on this early version of Slack in hopes of becoming profitable.

As a communication tool, the core of Slack's success comes from the ability to create community and facilitate productive social interaction, some of the core strengths of Butterfield's earlier products. Indeed, many of the learnings from Butterfield's MMOs were directly translatable, and Slack's basis in MMO games actually contributed to the success of the messaging app's core "freemium" business model.

---

[5] https://blog.asmartbear.com/freemium.html
[6] www.inc.com/business-insider/behind-the-rise-of-stewart-butterfield-and-slack.html
[7] www12.statcan.gc.ca/census-recensement/2016/dp-pd/prof/details/page.cfm?Lang=E&Geo1=DPL&Code1=590021&Geo2=PR&Code2=48&Data=Count&SearchText=Lund&SearchType=Begins&SearchPR=01&B1=All
[8] https://mastersofscale.com/stewart-butterfield-the-big-pivot/

Introduction to Freemium

To understand Slack's success with freemium, we must first understand the concept itself. Freemium is a portmanteau, combining the words "free" and "premium." It describes a common pricing strategy where a product is offered free of charge but additional features, upgrades, or expansions are offered for an additional price. This business model was initially used in the 1980s and is popular across industries today.[9] Across the Apple App Store and Google Play, around 88% of apps are either free or freemium.[10] While Slack exemplifies the freemium model in enterprise software, many companies across different industries use a freemium model to attract initial users and convert a dedicated subset into paying customers.

Successful freemium products are built on the concept of Newtonian engagement. This phenomenon, named after Newton's first law of motion (colloquially, an object in motion will stay in motion and an object at rest will stay at rest unless acted upon by an external force), states that "an engaged player of freemium service will remain engaged until acted upon by an outside force." For a freemium service to be successful, users need to be highly engaged with the product because the more attached they are to a product, the more willing they are to pay. Thus, the path to freemium success begins with building a strong and sticky product that your users will love.

Building a Strong Core Product

A successful freemium product is first and foremost a valuable product, and there are many ways Slack is able to bring unique value to its users. In particular, Slack is able to successfully execute in three main categories that are important for capturing customers. These are

1. Centralization
2. Customer focus
3. Delightful branding

Centralization

One of the main value propositions of Slack over email is that information is by default public. For background, Slack is built on the concept of channels (a messaging board meant a certain topic) which allows team members who join the channel to communicate in a shared chat. Whereas email depends on long

---

[9] www.feedough.com/freemium-business-model/
[10] www.statista.com/statistics/263797/number-of-applications-for-mobile-phones/

chains that are invisible unless you're cc'd, Slack's information is highly searchable, even for channels you aren't in. The persistent chat messages allow people to find necessary information even months later. Furthermore, Slack integrates with many third-party apps such as Google Drive, Dropbox, Google Calendar, and more, helping to centralize communication across platforms and eliminating the context switching that occurs when working across multiple programs. Their focus on centralization helps create a streamlined user experience that people love.

Customer Focus

While many of its features help provide a productive experience, the reason Slack was able to focus on building the *right* features is because of its intense focus on the user experience. This is not surprising because Slack's very first user was indeed Slack itself. From the very beginning, the Slack creators were invested in creating a high-quality user experience—for their own benefit. As they began to grow outward, their focus on user experience became even stronger. For example, before being available to the public, Slack was first released through a "preview version." A select number of companies were chosen to test out the platform so that the team could fix any major bugs, identify key features, and learn how to scale. The team was unwilling to release a product that had not withstood legitimate testing, and after the preview release, they addressed the feedback they received and then released it to the public. This strategy truly exemplifies their user focus because of their insistence on releasing only a fully tested and validated product. As an added bonus, this soft launch helped them acquire user testimonials and press coverage even before an official widespread release, giving them the credibility and exposure to acquire even more customers. To this day, the customer experience is one of Slack's main priorities.

Delightful Branding

Because of their work on Glitch, members of the Slack team knew how to make normally boring and time-consuming tasks in a video game, such as breaking rocks, into a highly enjoyable activity and were able to apply similar strategies to incentivize menial work tasks with Slack. One of the ways they did this was through their highly differentiated branding. While most enterprise software at the time displayed muted colors and a very sterile, professional environment, Slack's branding is bright and cheerful—even playful. Its generous usage of stickers and GIFs (animated images) further demonstrate Slack's focus on making communication enjoyable, showing Butterfield's understanding of what motivates social interaction.

Achieving Your Freemium Dreams

While the approaches above help shed light on Slack's success with users, we also want to understand how Slack was able to convert so many of them from free accounts to paying customers. One of the shocking things about all the features mentioned above is that they are even available with Slack's free version. In fact, every core feature of Slack is available through the free pricing plan. This is exactly how Slack manages to produce such high conversion rates: they make it easy to enter, hard to leave, and habit-forming while you're there. With this extremely sticky product, once you're stuck, you begin to pay.

Easy to Enter

Right from the get-go, Slack provides an extremely frictionless onboarding experience. In fact, you can begin using Slack in just five clicks. They also make it extremely easy for you to invite friends or coworkers to the platform, a highly important feature for any communication platform. Through a seamless process, any user can quickly create a team on Slack.

As mentioned previously, one does not have to pay to play with Slack. One of the hallmarks of the company is its unique approach to fair pricing. If your organization uses one of the paid versions of Slack, you only end up paying for the number of active accounts, not the entire size of the organization.[11] Thus, payment becomes very low stakes—if you don't end up using it, you don't end up paying for it.

Habit-Forming

Though easy to enter, the experience of using Slack is also fulfilling in itself. While highly focused on providing valuable features for users, Slack also creates functionality that is able to hook users into continued usage. Habits are formed through the cycle of cue, craving, action, and reward. Messaging apps like Slack are particularly well adapted to this framework because message notifications act as a cue and spark your attention, as demonstrated by the plethora of notifications that are turned on by default.[12] This elicits a craving to see what the message is about and respond to it. Once you click the notification (the action), your curiosity is satisfied, and you are able to respond with your own message (the reward). Through this cycle, a habit begins to form, and using Slack has become ingrained in your actions. Even more, you can begin to crave using the product.

---

[11] https://slack.com/help/articles/218915077
[12] https://slack.com/help/articles/201355156-guide-to-desktop-notifications

Hard to Leave

Slack's partnerships and integrations help create a centralized ecosystem that makes a more diffused system look horrendous. Even if other platforms offer similarly streamlined integrations, once you have all of your accounts tied to a particular platform, it takes a lot of effort to transfer all your information to a new system. Therefore, Slack's integrations make it an essential part of people's workflow, creating a hard-to-leave platform.

Furthermore, Slack relies heavily on a network effect. This means that its services become more and more valuable as more people are on the platform. If your entire company is on Slack, it becomes very difficult for a single user to jump ship and use something else. This helps create an immense amount of friction for anyone who tries to leave and also creates great defensibility for the product. Even better, make a platform that no one will want to leave. Though features can be copied, a community cannot be.

Freemium Conversion

By making it easy for users to join and hard to leave, Slack has an amazing opportunity to monetize, and they are very intentional about their strategy. The free account comes with the large majority of features; thus, any limitations of using the free account do not immediately affect the user; only once Slack becomes an integral part of their workflow do users begin to see the benefits of an upgrade. By then you're already hooked into their ecosystem, and it is difficult to leave. The main limitations of the free account are a capped number of searchable messages, capped number of integrations, and a lower amount of file storage. These do not begin to affect a user until after the product has been consistently used therefore allowing a user to comfortably use the free version until Slack becomes critical to their communication abilities. At this point, paying for premium features seems like the best solution; Slack has already become too important in their day-to-day activities.

Overall, Slack monetization strategy only works because of the great product it has built. Through a laser focus on user experience, Slack is able to provide value for users as well as creating a truly sticky product. Once the platform becomes integral to a user's workflow, they are able to convert free users to paying evangelists. With their widespread success and adoption, Slack has created a product that can truly revolutionize workplace communication.

Takeaways: Slack has experienced tremendous success as an enterprise software platform, especially with its high conversion rate. In order to create a successful product, they focused on making a centralized product that is easy to join, habit-forming, and hard to leave. Through this, their strategy was to monetize by offering storage upgrades that would be useful only once a user was entrenched in their ecosystem, leading to one of the highest freemium conversion rates in the industry.

Questions

1. While 30% is much higher than average conversion rates, do you think there is an upper limit on what conversion rates can be sustainably achieved?

2. Other companies such as Dropbox use a similar monetization strategy but have lower conversion rates. Why do you think that is?

3. Spotify is another company that has a high freemium conversion rate. What similar or different strategies do they use to do this?

4. How do expected freemium conversion rates differ across the industry? For example, would consumer entertainment products, such as gaming or video streaming, have a higher or lower average conversion rate than enterprise software?

Fake It 'Til You Make It

PayPal and Reddit's Unconventional Approach to Growth Hacking

When Max Levchin, Peter Thiel, and Luke Nosek cofounded PayPal, then named Confinity, in 1998, they were fighting against Goliath. PayPal, an online payment and money transfer system, was entering a deeply competitive market dominated by Billpoint, a joint venture between eBay and Wells Fargo. With such steep competition, PayPal needed to take a nontraditional approach. Where Billpoint had the credibility that came with the Wells Fargo brand and a focus on combating fraud, PayPal took a completely opposite approach, instead focusing on ease of use and using one other tactic—creating fake demand.

The Provenance of PayPal

Though calling this "lying" may seem harsh, that's exactly what PayPal was doing. They realized their product couldn't gain initial momentum by itself, an example of the classic chicken-and-the-egg problem (see Chapter 5 for more

© Griffin Kao, Jessica Hong, Michael Perusse and Weizhen Sheng 2020
G. Kao et al., *Turning Silicon into Gold*,
https://doi.org/10.1007/978-1-4842-5629-9_10

about this problem). As a payment platform, PayPal only provides value when people can pay other people. When the platform had zero users, there was no reason for people to join because there was no one to pay, thus the platform could not grow on its own. In order to jumpstart growth, the team had to take measures into their own hands. Moving into enemy territory, they identified eBay as a key way to increase awareness and distribute their product. They developed a key insight into the platform that they were able to exploit: the importance of email in winning auctions. With sellers as their targets, they created a bot that crawled eBay's listings and approached sellers with a deal. The bot would email sellers claiming to be buying items for charity but only wanting to pay using PayPal. If sellers accepted then, score!—a new user had been acquired. If not, at least they were exposed to the PayPal brand. Through this and other similar growth tactics, PayPal was able to get its product into the hands of users. After a while, its ease of use contributed to continued usage and further adoption, ultimately allowing PayPal to sell itself.

While this approach to faking demand might seem shady at first, PayPal provided value to users that made them continue to use the product. PayPal was not relying on this growth tactics to populate their entire platform. Instead, they focused on creating initial growth by channeling energy into and being intentional about finding target users and a key distribution platform. Through this approach, PayPal was able to seed the demand that allowed them to grow naturally. People liked PayPal because it was a good product, but there was work that had to be done to get people to see that. After the user base reached critical mass, network effects would help drive continued growth.

Reddit Rising

In 2005, the cofounders Steve Huffman and Alexis Ohanian of Reddit, an online link aggregation and discussion forum, faced a similar problem. Reddit was a platform that needed content to get users, and needed users to incentivize content production, but at its inception, it was completely devoid of users. Reddit had to figure out how to find early adopters that would allow their platform to grow. But how could they possibly create a community from scratch?

By faking it.

In the early days, the Reddit team channeled their energy into creating a ludicrous amount of posts under fake usernames in order to simulate popularity and usage. As reckless as this may seem, however, the team was very intentional about the type of content that they posted: they focused on posting exactly the high-quality type of content that the founders wanted to continue seeing on the platform. Full of content of a certain standard, the platform began to attract exactly the kind of people who were interested in high-quality content and, thus, would post similarly quality content. The fake content gave these real users a sense that the platform was actually alive, giving them the motivation to actively contribute to it themselves. Therefore, the initial fake posts helped create exactly the kind of culture that Huffman and Ohanian were hoping to have on Reddit. This initial push of fake content allowed Reddit to reach a point where all the content was indeed user generated only a couple months after launch.[1]

Fake Demand, Real Value

The key takeaway from these stories is not to throw away the entirety of one's moral code and dive into deception. On the contrary, these companies were wholly invested in creating great products for their users and thought deliberately about the best way to reach them. Creating fake demand or fake content is in no way scalable—but it doesn't need to be. They just had to get in the trenches and start working with users. After all, without customers there isn't anything to scale in the first place. As Paul Graham said, "All you need from a launch is some initial core of users. How well you're doing a few months later will depend more on how happy you made those users than how many there were of them."[2]

And posing as users gave them exactly that. By staging content, they were forced to experience what a user would experience: the flow of uploading content and commenting on threads firsthand. Thus, they gained the valuable perspectives of using their own product and a chance to interact with other users of the platform, allowing them to focus on what mattered—building a great user experience.

The lies only worked because there was substance behind them. PayPal had a key insight into distribution channels and where to insert themselves in the process, while also simply being a superior product whose ease of use made people happy to adopt the product. Reddit understood that posting fake content would only be useful if it helped create the right kind of community, and they were therefore highly intentional about the kind of content they posted and people they wanted to attract. The lies were solely a means to an end, to get the product in the hands of users and focus on making their experience great—and *that's* how you fake it 'til you make it.

Takeaways: Both PayPal and Reddit addressed the initial problem of finding users by creating their own fake demand. This short-term solution only worked because once real users joined their platforms, they found real value in the product.

---

[1] https://venturebeat.com/2012/06/22/reddit-fake-users/
[2] http://paulgraham.com/ds.html?viewfullsite=1

Questions

1. The stories in this chapter illustrate times when faking demand leads to legitimate user adoption. How might a company's efforts to fake demand backfire?

2. What other strategies could Reddit and PayPal have used to get their first users? How might these other methods compare to the method of faking users?

Extending the Lead

Online to Offline (O2O)

Amazon's Dominance On and Off the Web

As we increasingly transition our lives onto the Internet, the line between the online and offline worlds is blurring. Activities that were originally done purely in person, such as education or gambling, are now transitioning online; perhaps the most prevalent example is shopping, which takes place in a hybrid of physical retail locations and retailer web sites. Nowadays, many direct-to-consumer (D2C) companies even sell exclusively online. O2O is an acronym used in digital marketing that simultaneously stands for "online to offline" and "offline to online," describing how retail and e-commerce companies manage their customers both on the Internet and in the physical world. While the acronym has dual meanings, it is more commonly used to describe "online-to-offline" strategies that entice consumers in a digital environment to make purchases of goods/services from physical businesses.[1] Though there are numerous variations of the O2O model, such as online ads that drive

---

[1] www.investopedia.com/terms/o/onlinetooffline-commerce.asp

© Griffin Kao, Jessica Hong, Michael Perusse and Weizhen Sheng 2020
G. Kao et al., *Turning Silicon into Gold*,
https://doi.org/10.1007/978-1-4842-5629-9_11

consumers to visit physical stores, or QR code marketing efforts, one of the most interesting O2O trends is the opening of brick-and-mortar stores by e-commerce companies.[2]

- Warby Parker—eyewear
- Glossier—makeup
- Casper—mattresses
- ThirdLove—bras

There is a long list of e-commerce companies that started exclusively online and which have migrated, at least in part, to physical store locations. And, of course, this includes the most iconic example: Amazon.

Amazon and Disruptions in Retail

Amazon's humble beginnings as an online bookstore can be traced back to July 5, 1994, when it was founded in Jeff Bezos' garage.[3] Since then, the online bookstore has revolutionized retail, with several milestones along the way:

- In 1998, Amazon expanded beyond books to sell CDs and DVDs.[4]

- In 1999, Jeff Bezos was named Time magazine's Person of the Year and dubbed "the king of cybercommerce."[5]

- In 2005, Amazon introduced Amazon Prime,[6] which had over 100 million subscribers at the end of 2018 (for context, that's almost one-third of the US population of 320 million around the same time).[7]

- In 2007, Amazon started selling the Kindle, its first consumer electronic product.[8]

There are numerous other landmark events in Amazon's history—all of which are impressive and have contributed to the evolution of the company into the king of online retail in less than 25 years.

---

[2] www.brand-experts.com/digital-trends/e-commerce-trends/
[3] www.cnn.com/interactive/2018/10/business/amazon-history-timeline/index.html
[4] Ibid.
[5] Ibid.
[6] Ibid.
[7] www.businessinsider.com/amazon-prime-members-spend-more-money-sneaky-ways-2019-9
[8] Ibid.

Perhaps surprisingly, after achieving immense success on the Internet, this online behemoth began branching into physical retail.

Amazon's first physical bookstore, Amazon Books, opened in 2015. It acquired Whole Foods Market in 2017, which had 500 locations as of 2019.[9] It opened Amazon Go, a cashier-less grocery store, in 2018, and is planning to open 3,000 Amazon Go stores by 2021.[10] It has opened four "Amazon 4-Star" stores since 2018.[11] There are numerous Amazon Pop-Up shops across the United States.

Considering how much success the company has found online and how many of its brick-and-mortar competitors are struggling, it certainly seems strange that Amazon is pushing so strongly for an offline presence. In fact, the closing of a large number of North American brick-and-mortar retail stores in the 2010s has been dubbed the "retail apocalypse" (or "retailpocalypse"), with Sears, Toys "R" Us, and Gymboree being just a few of the victims.[12] This "death of retail" has been attributed to overexpansion of malls, aftereffects of the Great Recession, and importantly, changing consumer spending habits induced by the rise of e-commerce.[13] So, why then is Amazon looking to enter into the world of physical retail that it itself has been pushing to ruin?

Brick-and-Mortar and Prime

There might be more reasons than one would expect, and most are related to Prime. These physical store locations are a real-world embodiment of Amazon's retail ecosystem. When Amazon introduced Prime in 2005, it introduced the core of its multibillion-dollar consumer retail business. Not only does Amazon earn money from consumers signing up for Prime memberships, Amazon also incentivizes Prime members to shop *more* through the various perks that are offered (free shipping, TV streaming, etc.)—as a result, Prime members spend more than twice as much as non-Prime members on Amazon.[14] The physical stores are merely an extension of this; certain

---

[9] www.supermarketnews.com/retail-financial/amazon-adds-physical-retail-footprint-latest-go-store

[10] www.supermarketnews.com/retail-financial/what-amazon-s-brick-and-mortar-disruption-could-look

[11] www.seattletimes.com/business/amazon/amazon-opens-4-star-store-at-seattle-headquarters-as-online-giant-grows-physical-shopping-presence/

[12] www.cbinsights.com/research/retail-apocalypse-timeline-infographic/

[13] www.theatlantic.com/business/archive/2017/04/retail-meltdown-of-2017/522384/

[14] www.businessinsider.com/amazon-prime-members-spend-more-money-sneaky-ways-2019-9

products in the physical stores have deals and discounts exclusively for Prime members, which could yield a purchase from a Prime member who otherwise wouldn't have spent that money.[15]

These stores are also a huge marketing campaign in and of themselves. Many of them sell Amazon's various flagship products, such as Echo devices, Kindle e-readers, and more.[16] This drives sales of these Amazon products and also drives more Prime subscriptions. Amazon's various Alexa and Echo devices all offer Prime-exclusive perks, including skills like conveniently (re-)ordering products, sharing limited deals, showcasing pictures from your Prime Photos, and more. The real value of Echo comes through Prime subscriptions (a key driver in Amazon's business strategy, as discussed above!), and so Amazon stores are a brilliant way to bring more customers into its extremely effective customer loyalty program while driving sales of its consumer electronics.

Of course, physical stores also help Amazon forge stronger relationships with its customers. Although the ease and convenience of Amazon's online shopping experience are exceptional, it still doesn't replace an in-person shopping experience with face-to-face customer service and the feeling of satisfaction in browsing an immersive, physical space. Employees at Amazon Go, its cashier-less grocery store, are there to provide top-notch service and to answer any questions.[17] These human relationships are ultimately necessary for building strong, long-lasting customer relationships. Perhaps more importantly, these stores are also shaping customers' impressions of Amazon via the physical shopping experience. Amazon prides itself on its consumer centricity through its renowned two-day shipping and 1-Click ordering, and its cashier-less grocery stores take that shopping convenience into the real world. Its bookstores, "4-star" stores, and pop-up stores showcase its diversity of products and transform the act of shopping into a leisurely activity. These all echo the concept of a flagship store used by other retail chains, where the main goal isn't necessarily to make the maximum amount of sales but rather to immerse the customer in the brand and develop a long-term customer relation. Amazon recognizes this importance of physical immersion and has redoubled its efforts to maintain strong customer relationships.

Amazon's Foray into Groceries

Amazon has clearly spent much time and effort entering the world of physical retail, and we now turn our attention toward a very specific niche of physical retail: groceries. Amazon has been dabbling in the grocery business since 2007, though rather unsuccessfully, with its online food delivery service,

---

[15] http://nymag.com/intelligencer/2017/06/why-is-amazon-building-bookstores.html
[16] https://fortune.com/2017/04/28/5-reasons-amazon-physical-stores/
[17] www.destinationcrm.com/Articles/ReadArticle.aspx?ArticleID=129179

AmazonFresh.[18] Despite the company's massive size, grocery is a huge component of retail that Amazon hasn't quite captured yet. It's something that we all buy with high frequency on a weekly basis. If Amazon can capture sales in this area, it's a huge win for them.

In that sense, it absolutely makes sense that they're moving toward brick-and-mortar shops, which offer a sense of security for freshness of products that simply can't be provided online (yet). The company took their first step toward this by purchasing 421 Whole Foods stores in June 2017.[19] Not only do the stores provide the "freshness" factor beneficial for groceries, they also extend Amazon's network of supply depots and can be combined with Amazon's already-impressive logistics and delivery system to push the grocery delivery business to higher heights.

O2O and D2C

Although not all of these reasons for moving from online to offline are applicable to other D2C e-commerce businesses, the core concept is there: it is vital for companies to integrate their online and offline strategies. In today's growing digital world, businesses need to grow alongside the consumers and acknowledge the intertwining of our online and offline worlds. They need to strategically think about how they can integrate both to create the best customer experience. There exists a symbiotic relationship, where their online presence may drive awareness, while their offline stores may provide immersive experiences and a human sense of trust that reinforces their online image.

Companies need to find a balance for their online platforms to coexist with the real world. This will look different for different companies, but we can all find guidance in Amazon's strategic use of its Prime program and how it has managed to incorporate the customer loyalty program both on the Internet and in the real world in order to further establish its customer base and achieve vertical integration.

Takeaways: In today's day and age, companies should effectively integrate their business both online and offline. Amazon has effectively brought their online business into the physical world via brick-and-mortar shops that branch out their customer loyalty program, offer Amazon's flagship products, enhance the customer experience, and help expand their supply network in a new product space (groceries). D2C companies opening retail stores can look toward Amazon for inspiration on how to open delightful yet successful brick-and-mortar shops.

---

[18] http://nymag.com/intelligencer/2017/06/why-did-amazon-buy-whole-foods.html
[19] Ibid.

Questions

1. What products would you only purchase offline? Which would you only purchase online? Do you have different expectations for the shopping experience in these cases?

2. Have you visited any of Amazon's physical stores? How did you find the experience, and do you think it changed your view of Amazon as a company?

3. Since D2C companies often sell directly online, what do you think is the importance of having a "flashy" flagship store? What does the physical store say about the store's brand and image?

The Last Mile Problem

The Biggest Thorn in Amazon's Side

Slow and steady, the box-shaped vehicle chugs along on a well-paved sidewalk, passing well-manicured suburban homes, one after the other. It's the size of a large dog and doesn't look like it could fit any normal-sized human, but on six rugged wheels, it looks determined to overcome any obstacles that could get in its way. At the same time, the baby blue shell and the Amazon smiley face logo slapped onto its side give it the appearance of a friendly, loyal companion—Fido reincarnated to match our increasingly tech-dependent world. While its purpose is unclear, the vehicle seems to suggest an impending utopia where human and automaton coexist in peaceful harmony. When it finally stops at a beautiful house and immediately lifts open its top to reveal a package marked with the unmistakable blue and black Amazon Prime packaging, the mission of the autonomous vehicle is also revealed. The home's resident meanders out from the front entrance to retrieve her package, smiling at how quickly and conveniently her Amazon ordered has been delivered.

The vehicle is called Amazon Scout, the latest in a line of solutions being explored by Amazon to tackle the thorny last mile delivery problem—a problem that has become more and more important to address as consumers turn to the Internet for their shopping needs. These online shoppers increasingly

© Griffin Kao, Jessica Hong, Michael Perusse and Weizhen Sheng 2020
G. Kao et al., *Turning Silicon into Gold*,
https://doi.org/10.1007/978-1-4842-5629-9_12

demand fast shipping at a low cost, having grown accustomed to reasonably priced same-day or two-day shipping through Amazon Prime or similar programs. Unfortunately, the issue lies in the disproportionate cost of the last leg of the delivery: getting the package from some high-capacity local freight station or port to the final destination, the customer's doorstep. The cost of this part of the journey, while just a tiny fraction of the distance covered by the package, accounts for 53% of the total shipping cost.[1]

Amazon and Last Mile Delivery

Although last mile delivery is a predicament to the online retail industry as a whole, Amazon's 49% market share of e-commerce means the retail giant ships a lot more packages than anyone else.[2] In particular, the more than four billion parcels they shipped in 2018 cost the company an estimated $12B for outbound delivery (from Amazon warehouses or shipping centers to customers).[3] This figure represents over 5% of their overall operating expenses for the year.

When we consider growth trends, these costs become even more concerning for Amazon. The company's revenue skyrocketed between 2004 and 2017, growing by 240 percent from $74B to $177B, and yet shipping costs managed to outpace those revenue gains.[4] In this time span, shipping costs grew by 330 percent from $6.6B to $21.7B.[5] One explanation for such a jump in shipping costs is that this metric has been inflated in recent years by investments Amazon has made into its logistics network—in infrastructure such as warehouses, sortation and fulfilment centers, delivery vehicles, and the personnel to run all these pieces—to reduce external dependency and minimize future expenses. However, despite these investments (expanded upon later), Amazon continues to lack an efficient and scalable solution to last mile delivery, and until they have one, shipping costs augmented by the growing number of orders will continue to eat away at their profit.

Perhaps more importantly, the last mile is personal to Amazon. Since conception, the company has been obsessed with vertical integration (in which a company owns multiple parts of the supply chain). Amazon has expanded into other areas of the production vertical and acquired adjacent parts of the supply chain with unprecedented aggression. With benefits such as reducing production

---

[1] www.businessinsider.com/last-mile-delivery-shipping-explained
[2] https://techcrunch.com/2018/07/13/amazons-share-of-the-us-e-commerce-market-is-now-49-or-5-of-all-retail-spend/
[3] www.savethepostoffice.com/an-amazon-puzzle-how-many-parcels-does-it-ship-how-much-does-it-cost-and-who-delivers-what-share/
[4] www.statista.com/topics/846/amazon/
[5] www.savethepostoffice.com/an-amazon-puzzle-how-many-parcels-does-it-ship-how-much-does-it-cost-and-who-delivers-what-share/

costs, creating new distribution channels, and capturing upstream and down-stream profits, Amazon has taken vertical integration to the next level by making it not only a core part of their business strategy but an integral part of their company identity. Their mission is total control over the end-to-end customer retail experience, because a loss of control is risky—the customer experience could fail in the final stages of a package's journey. Thus, the company cannot reconcile their own success with a continued reliance on third-party shipping services like UPS, FedEx, the US Postal Service, and regional carriers to handle delivering packages to customers' doorsteps.

Amazon's Solutions

It's no wonder that Amazon has prioritized finding a solution to the last mile problem. Indeed, they've created an entire team devoted to the pain point. Called the Last Mile team, the division was looking to hire for 200 full-time roles as of October 2019, including various tech roles such as software engineers, product managers, and machine learning scientists.[6] In addition to Scout, Amazon has explored other options, including letting users pick up their own packages through Amazon Locker (a self-service pickup kiosk), using drone delivery through Prime Air, and outsourcing last mile delivery to willing entrepreneurs through the Amazon Delivery Service Partners program. The Partners program attempts to apply the increasingly popular crowdsourcing solution (see Chapter 24) by recruiting enterprising individuals to start their own companies to deliver Amazon orders, claiming they could earn up to $300,000 in annual profit from operating up to 40 delivery vehicles.

Unfortunately, each solution seems to have some not-so-trivial problem. Amazon Lockers require large amounts of space only available in the brick-and-mortar stores that are being killed out by Amazon itself (see Chapter 11 for a more nuanced dive into Amazon's relation with brick-and-mortar). Drones have to deal with poor weather, short ranges, and government regulation, among other issues. And crowdsourcing introduces the issue of having to incentivize actual humans at a scale that matches rising order volume. Many other players in the space, like UberRush, Deliv, and Sidecar, have attempted and thus far failed to solve the last mile with similar strategies.

Implications of a Solution

Amazon Scout or autonomous grounded delivery devices may be the most promising option of the bunch. The Scout vehicles are being touted as "a new, fully-electric delivery system" that can "safely and efficiently navigate around

---

[6] www.amazon.jobs/en/teams/last-mile

pets, pedestrians and anything else in their path" by Amazon's blog Day One.[7] The initiative could mean Amazon is one step closer to figuring out how to deliver packages more quickly and more cheaply at scale. But even if none of the current solutions or even a combination of the solutions are remotely close to solving the last mile problem, the money and brand on the line for Amazon (as well as the resources available to the company) mean the tech powerhouse will come up with a solution sooner or later.

When that time comes, the cost of online shopping will drop rapidly as shipping costs are reduced, and the growth of the e-commerce industry will accelerate. But not only will physical stores feel the pressure as online shopping becomes a much more appealing alternative (we could see many more closures), other online retailers could also suffer if Amazon's able to patent or monopolize the operation of their last mile solution. Moreover, shipping services like UPS would not only lose the considerable business that Amazon gives them, but they could see themselves locked in a losing shipping battle with the giant—another industry added to the long list of ones Amazon has already disrupted. The only winner in this scenario besides Amazon might be the customers who get to enjoy lower prices. Without a doubt, the stakes are sky high, and the last mile will prove to be a critical battleground for Amazon and other major players in the retail and shipping industries.

Takeaways: Finding a solution to the last mile problem is a priority for Amazon for economic and identity reasons. So if/when Amazon finds a solution, retailers and shipping services will suffer, but customers will benefit. More generally, Amazon's vertical integration strategy has proven fruitful in reducing costs and increasing control over the customer experience, proving to be a useful strategy for mature tech companies.

Questions

1. What personal stake do you have in the resolution of the last mile problem? Are you in the retail industry? Are you in the shipping industry? Are you an online shopper?

2. What is likely to be the response by brick-and-mortar retailers to greatly reduced shipping prices? How can brick and mortar continue to compete with online retailers as shipping costs reduce?

3. The US Postal Service, a semi-independent federal agency, reported a revenue of $69.6 billion for the fiscal year 2017. How would the US government respond if Amazon displaces US Postal Service as a shipping service?

---

[7] https://blog.aboutamazon.com/transportation/meet-scout

13

Tech IPO Flops

Why It Seems Like Tech Companies Fall Hardest in the Public Market

In the Spring of 2019, the world's eyes were on Uber as the ride-hailing giant geared up to debut on the open market at $45 a share. Investors, on the edge of their seats, were simultaneously nervous and hopeful that the shares offered in the initial public offering (IPO) would experience that first day jump in value caused by a rush of individual investors trying to get in on the action. They hoped that public perception would match that of the private investors that had gorged Uber with so much venture capital in the years prior. But alas, it was not meant to be—a debacle erupted as the value of the company's shares plummeted from the opening price on that first day.

Nearly every major news outlet paying attention to the stock market, from Bloomberg to Fortune.com to *The New York Times*, ran articles reporting on Uber's massive failure in going public. One *Vanity Fair* article was headlined "Uber's Colossal I.P.O. Flop May be the Worst Ever on Wall Street," bluntly capturing the severity of the bust.[1] By comparison, the average return on 62 IPOs in the second quarter of 2019 was 30%, with Beyond Meat (the company selling plant-based meat) up around 542% at the end of the quarter.[2]

---

[1] www.vanityfair.com/news/2019/05/ubers-colossal-ipo-flop-may-be-worst-ever-on-wall-street
[2] www.cnbc.com/2019/06/28/ipos-have-their-best-quarter-in-years-in-terms-of-performance-and-capital-raised.html

© Griffin Kao, Jessica Hong, Michael Perusse and Weizhen Sheng 2020
G. Kao et al., *Turning Silicon into Gold*,
https://doi.org/10.1007/978-1-4842-5629-9_13

However, Uber wasn't alone in its plight. Lyft, IPOing soon after in Q2 2019, also flopped, declining 12% in share value on the first day. In fact, the two ride-sharing companies joined a long list of tech IPO underperformers, all trading below the initial share price after entering the public market—a list headed by the likes of Groupon, Pets.com, Zynga, Pandora, and Facebook, one of the most notable flops of all time as its shares had fallen to less than half of the IPO price at one point.

What Happens in an IPO

So what causes all of these duds? To begin to understand what's behind the IPO flops,[3] it's critical to examine how IPOs work. At a high level, IPOs occur when a private company needs to raise capital because it can no longer self-finance with operating cash flow. The company can choose to do so through a public investment called an IPO (where ordinary people like us can essentially pay the company for a small share), rather than incurring debt or attempting to find private funding. All companies listed on the stock exchange have gone through the IPO process. In the process, a company will typically hire an underwriter, usually an investment bank, to facilitate the public offering by buying shares from the company and selling them to the public (in a bought deal) or by selling the shares directly to institutional clients/the investing public (in a best effort deal). The underwriter will set some "offering price" tailored to the perceived demand and the volume of capital the company would like to raise. Usually only the institutional investors can buy shares at the offering price. Conversely, the "opening price" is based on supply and demand and is the price of a share on the day of the IPO. Shares opening below the offering price occurs if, in the first day, more sell orders (requests to sell shares) are submitted by investors than buy orders (requests to buy shares).

The Profitless Prosperity Model

In this context, an IPO flop occurs when the public values shares less than the company selling the shares or institutional investors do. One theory as to why this may seem to happen more often for tech companies points to an increasingly prominent business model used by those fast-growing tech start-ups: the profitless prosperity model. In the profitless prosperity model, companies focus less on profits and more on rapidly accumulating market share, typically in highly competitive industries. Companies that successfully implement this approach can point to massive revenue generated by an explosive user base but oftentimes coupled with substantial losses, like Tesla's -$408M net income in 2018.[4]

---

[3] Note that we use "flop" loosely—we use it to mean a general failure.
[4] www.macrotrends.net/stocks/charts/TSLA/tesla/net-income

If you think of a major tech company, like Uber, Lyft, and nearly every social media platform, there's a good chance that they either currently use the profitless prosperity model or once did. And they have good reason to since Amazon was once a profitless pioneer, reporting a net loss of $124.5M the year after it IPOed.[5] In contrast, the company made $10.073B two decades later in 2018 with the company's shares selling at more than 120,000% their initial IPO value. The idea is that once a company obtains market leadership, they can begin to reduce costs or improve their monetization strategy.

While private investors, particularly venture capital firms providing seed funding, have a voracious appetite for high-potential, high-risk bets, individual investors have a lower tolerance for risk. Thus, revenue growth in the face of continuing loss could make buying shares in a tech IPO without seeing some performance on the open market less appealing—in this case, there might be some net loss threshold that causes an IPO to flop.

Longer Incubation Periods

Contributing to a potential disillusion with the profitless model is that the period of time a tech start-up will remain private from conception to IPO is being increasingly inflated by the availability of private capital. *The New York Times* has pointed out that when Netscape and Yahoo went public in the 1990s, neither had existed for more than 3 years, but Lyft was 7 years old, and other companies IPOing in 2019 like Uber, Pinterest, and Zoom were all older than that.[6] This means many of these companies have had more time to develop sustainable business models while gobbling up private money—making those that haven't stick out like a sore thumb. For example, in the year before IPOing, Zoom had a net income of $7.6M (a stark contrast to Uber's Q4 losses of $768M in 2018) and saw a massive $30 jump from offering price to opening price per share.[7] In addition, Zoom shares grew in value to $95 by August 2019.

Market Sensitivity to News and Tech Scrutiny

These ideas underscore a market sensitivity to company news—which could be anything from a bleak financial report to a Twitter scandal involving a federal lawsuit. Here, the profitless prosperity model comes into play yet again—tech companies, often consumer facing, are very much in the public eye

---

[5] www.macrotrends.net/stocks/charts/AMZN/amazon/net-income
[6] www.nytimes.com/interactive/2019/05/09/business/dealbook/tech-ipos-uber.html
[7] www.cnbc.com/2019/03/22/video-conferencing-company-zoom-files-to-go-public-is-profitable.html

because they've prioritized market share and have acquired astronomical consumer bases (Uber has 110M active users worldwide as of 2019).[8] This means the heightened scrutiny surrounding tech companies may make their IPOs particularly volatile.

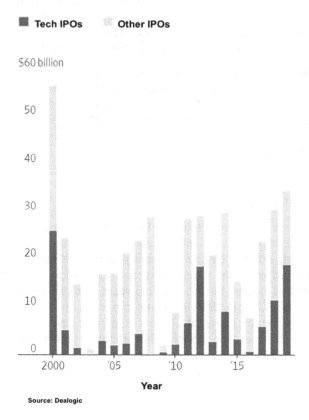

Figure 13-1. Tech IPOs as a Fraction of Total IPO Value

But this also brings us to an alternate explanation—the sheer representation of tech IPOs out of all IPOs just makes it *seem* like they flop often and hardest. In the first half of 2019, tech IPOs raised $18.8B which was 54.6% of the total capital raising through public offerings (see Figure 13-1).[9] Although 2019 might be a slight outlier, the 21st century has been sprinkled with blockbuster tech IPOs that often draw extensive media coverage. So it's highly possible that

---

[8] www.statista.com/statistics/833743/us-users-ride-sharing-services/
[9] www.wsj.com/articles/race-to-public-markets-continues-despite-uber-lyft-flops-11560684604

tech companies don't flop at a higher rate or at a higher degree than any other type of company—their exposure to the general population just makes it seem like they do.

Although it's impossible to know with certainty if tech IPOs actually perform poorly when compared to other companies or if so, why that happens—a key insight to pull out from all of this is that IPOs, like the market in general, are unpredictable. However, earning reports and press coverage and venture capital trends may provide early indicators for the ones that do flop.

Takeaways: The proliferation of the profitless prosperity model, longer periods spent as private companies, market sensitivity to news, and overrepresentation of tech companies in the news/IPOs may all contribute to why tech IPOs seem to flop hardest.

Questions

1. From the standpoint of an individual investor, what is your tolerance for risk?

2. Do you care about net income when considering investing in a tech stock? And if so, what threshold of profit means you'll buy a stock?

3. If you plan on starting a company, what are some better alternatives to the profitless prosperity model in your space? Or, what are some low overhead solutions that could lead to rapid net profit?

Apples to Apples

How Market Cannibalization Makes Apple One of Its Biggest Competitors

If you currently own an iPhone, you've probably owned *many* iPhones. Whether it's the intuitive design, sleek body, or ability to send blue texts that keeps you hooked, you've chosen Apple for your phone—and maybe your laptop, desktop, tablet, and watch, too. But in many Apple users' experiences with the company, there comes a tipping point where you decide that the next, shiniest model of the *iSomething 20* is simply not worth the upgrade; your old *iSomething 19* will do. In fact, your iSomething 19 is still *so good* at its job that you don't see the need to buy any new Apple products for quite some time.

This scenario is a common and troubling one for consumer technology companies such as Apple, who find themselves as one of their own biggest competitors while extending their various product lines like iPhone. But how can a company be its own competitor?

Understanding Market Cannibalization

Let's start by examining the scenario of Apple releasing a new iPad. By doing so, Apple opens up an opportunity to make some sales and gain more market share in tablets with a shiny new product, but it simultaneously hurts itself in two ways.

© Griffin Kao, Jessica Hong, Michael Perusse and Weizhen Sheng 2020
G. Kao et al., *Turning Silicon into Gold*,
https://doi.org/10.1007/978-1-4842-5629-9_14

First, older iPad models will see their sales drop off at a tremendous rate. While third-party markets such as eBay will see an increase in the sale and transfer of these older products (to be replaced by newer models), Apple does not directly benefit from this. Thus, by releasing a new product, Apple hurts itself and its past investments by decreasing sales of past iPad models. Second, when Apple chooses to invest in and release a new model, that product must compete against its older models that users currently own. Users have to evaluate the trade-off between the new model and its cost vs. their old model, effectively pitting Apple products against themselves. This competition extends beyond a given product line as well; for instance, new iPad models have to compete with new and existing iPhones and Macs as well, as all three Apple product lines offer similar features and functionalities.

This phenomenon is known as **market cannibalization**, defined as a reduction in sales volume, sales revenue, or market share of one product as a result of the introduction of a new product by the same producer.[1] Apple "cannibalizes" its market every time it releases a new model of a preexisting product. This contrasts with releasing new product lines—such as the introduction of the Apple Watch in 2015—where Apple is entering a new market that doesn't have any "players" yet.

Why does Apple continuously choose to do this? In short, (1) they feel the need to innovate to retain and excite their users, (2) new releases get a lot of press (free advertising!), and (3) moving away from the yearly cadence of new releases would be unexpected and disappointing, adding volatility to their stock. But market cannibalization is certainly a problem that Apple is becoming increasingly aware of, as Apple saw its iPhone sales drop for the first time ever in 2018 and continue to drop by 17% in 2019.[2] And when the iPhone makes up nearly 70% of the technology titan's revenue, this is a horrifying reality.[3]

For insights into Apple's history with balancing market cannibalization and innovation, let's take a closer look at the timeline of their most famous product—the mighty iPhone.

History of iPhone Models (2007–2019)

It's January 9, 2007, and Steve Jobs is standing on a simple stage in his iconic black turtleneck with a simple slideshow behind him.

"Today, we're introducing three revolutionary products..." he says with a smile.

---

[1] https://en.wikipedia.org/wiki/Cannibalization_(marketing)
[2] www.bbc.com/news/business-48110709
[3] www.imd.org/research-knowledge/articles/apples-dwindling-sales-show-importance-of-self-cannibalization/

"The first one is a widescreen iPod with touch controls. The second is a revolutionary mobile phone. And the third is a breakthrough Internet communications device." The audience begins to hum as the world prepares to hear about Apple's next greatest inventions.

"These are not three separate devices," Steve adds on. "This is one device, and we are calling it, iPhone."

In this speech, Jobs is clearly aware of market cannibalization that the iPhone would breed toward the iPod. This new device would be an iPod and so much more, and today we can see the effects of this as iPod sales have drastically dwindled to a fraction of what they once were.[4]

To better understand what happened from the initial iPhone launch to the 2018/2019 decline in sales, let's look at an overview of all of the iPhone models up until 2019 in Table 14-1.[5]

Table 14-1. iPhone Models 2007–2019

Release year	Model	Starting price	New features (excluding general trends in iOS updates and most hardware updates)
2007	iPhone 1st Generation	$399	Touch screen keyboard, larger screen
2008	iPhone 3G	$199	App Store, GPS, white color option
2009	iPhone 3Gs	$199	3-megapixel camera and video capture
2010	iPhone 4	$199	Front-facing camera (#Selfie)
2011	iPhone 4S	$199	iMessage, 1080p photo and video, Siri
2012	iPhone 5	$199	Lightning cable, LTE-4G speeds
2013	iPhone 5S	$199	Touch ID
2013	iPhone 5C	$99	Plastic model, new colors
2014	iPhone 6 and 6 Plus	$649 and $749	Larger size options
2015	iPhone 6S and 6S Plus	$649 and $749	3D touch, new colors
2016	iPhone SE	$399	iPhone 5 size with iPhone 6 internals
2016	iPhone 7 and 7 Plus	$649 and $769	Portrait mode, no headphone jack, haptic home button

(continued)

---

[4] www.lifewire.com/number-of-ipods-sold-all-time-1999515
[5] www.theiphonewiki.com/wiki/List_of_iPhones

Table 14-1. *(continued)*

Release year	Model	Starting price	New features (excluding general trends in iOS updates and most hardware updates)
2017	iPhone 8 and 8 Plus	$699 and $799	No major feature additions
2017	iPhone X	$999	Face ID, OLED display, notched screen
2018	iPhone XS and XS Max	$999 and $1,099	Larger size
2018	iPhone XR	$749	Back to LCD screen (cheaper option)
2019	iPhone 11 and 11 Pro/Pro Max	$699, $999/$1099	Ultrawide camera, liquid retina display, night mode

Some takeaways from this chart:

1. Even years have traditionally launched with more substantial updates (except possible the iPhone in 2017).

2. 2014 marked the first year with no "affordable" starting price; prior years had $199 starting options for phones with small amounts of memory.

3. Size changes become the traditional *biggest* update to be expected starting in 2014.

iPhone Sales Problem

At a glance, the sales crash in 2018 seems to make sense; users who had older iPhone models in 2018 (such as the 2015 iPhone 6S Plus) had practically no reason to update for new features, sizes, or iOS updates (since 2015 phones could run the latest version). And combined with controversial changes such as the removal of the headphone jack, the crash makes even more sense. The 2018 market for iPhone users had been saturated, in that fewer people were buying their first iPhones in 2018 and 2018 iPhone models such as the iPhone XS experienced an unusual amount of competition with older models. iPhone sales data, combined with the fact that iPhone sales consistently make up about half of Apple's revenue, support the hypothesis that the company had finally been beaten by versions of its past self.[6]

---

[6] www.statista.com/statistics/263401/global-apple-iphone-sales-since-3rd-quarter-2007/

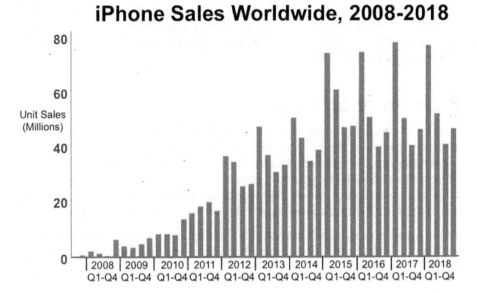

Figure 14-1. iPhone Sales 2007–2018[7]

Planned Obsolescence Backlash

Decline in iPhone sales might also have been attributed to a growing frustration with Apple's lethargic, gradual improvements to its products. iPhone memes surfaced in popular culture to parody the way in which the tiniest change to the iPhone would be lauded by Apple as technological genius, particularly boasts about minor adjustments to the phone's size.[8] Furthermore, users began to suspect that Apple had been making older iPhone models slower and less usable via software updates in an effort to boost new iPhone sales, culminating in a 2017 class action lawsuit.[9]

This strategy of gradually improving products at minor rates (and holding back more substantial improvements for later models) in addition to intentionally devaluing older products is known as **planned obsolescence**. Defined as "a policy of producing consumer goods that rapidly become obsolete and so require replacing, achieved by frequent changes in design, termination of the supply of spare parts, and the use of nondurable materials," Apple checks the box in many of these categories. By gradually adding one or two important

---

[7] www.statista.com/statistics/253649/iphone-revenue-as-share-of-apples-total-revenue/
[8] https://knowyourmeme.com/memes/subcultures/iphone
[9] www.forbes.com/sites/adamsarhan/2017/12/22/planned-obsolescence-apple-is-not-the-only-culprit/#cfe25003cf24

features per update, Apple is able to provide a more steady, stable flow of sales while saving some innovative yet potentially riskier features for later models. As iPhone sales dropped in 2018, it seems like the updates to the iPhone were deemed too gradual to be worth the cost.

Problem or Strategy?

Despite the negative impacts surrounding market cannibalization, Apple has long been aware of the phenomenon. When asked about how the launch of the iPhone would impact iPod sales, Steve Jobs is reported to have said that "if you don't cannibalize yourself, someone else will."[10] Using the exact verbiage of *cannibalization* shows that Apple has intentionally made bets against its past self in an effort to innovate, capture, and retain market share.

When asked about cannibalization, Phil Schiller, an executive at Apple, made it clear that Jobs' spirit of self-cannibalization was alive and well, especially across product lines:

> [Cannibalization] is not a danger, it's almost by design. You need each of these products to try to fight for their space, their time with you. The iPhone has to become so great that you don't know why you want an iPad. The iPad has to be so great that you don't know why you want the notebook. The notebook has to be so great, you don't know why you want a desktop. Each one's job is to compete with the other ones.[11]

—Phil Schiller, Apple Senior Vice President

Apple has been able to take risks at market cannibalization while simultaneously remaining innovative enough to grow its market share and retain users. By designing their products to synergize, increase usability, and focus on niche uses for each product, they've been able to grow the number of Apple devices people interact with in any given day. Whether the increasing number of devices or iterations on each device will continue to be a successful strategy remains to be seen.

Takeaways: Market cannibalization is an occurrence where a company finds itself competing with itself across two or more products. Cannibalization is a risk, but it can pay off if it means having excellent products that both attract more market share overall that either one would do separately. In the case of Apple, it has been a strategy.

---

[10] www.imd.org/research-knowledge/articles/apples-dwindling-sales-show-importance-of-self-cannibalization/

[11] www.cbsnews.com/news/60-minutes-apple-tim-cook-charlie-rose/&sa=D&ust=1573461692343000&usg=AFQjCNEya4hbE6NOi15CkAEiamOCFhNF2w

Questions

1. To what extent do you think Apple is guilty of planned obsolescence? Is it conspiratorial to think that Apple intentionally withholds feature updates or throttles older phones to maximize new sales?

2. What other companies do you think experience market cannibalization? Can you think of any that are software-specific?

3. Do you think Apple will create a new product as revolutionary as the iPhone, or will we just see iterations for the next 50 years?

Memes As Marketing

How Wendy's and WeChat Went Viral

How many fast-food restaurants have released a chart-topping mixtape? The answer is exactly one—Wendy's. Their mixtape, *We Beefin?*, tears down competitors like McDonald's and Burger King while also managing to reach the number one spot on Spotify's Global Viral 50[1] and garnering almost 800 million impressions.[2] But, Wendy's isn't just a one-hit wonder. Their Twitter account is also a hit, managing to achieve a 113% jump in Twitter mentions. Through consistently clever tweets, Wendy's brands themselves as "the sassy friend you want to go to lunch with". These marketing wins have also demonstrated a major business impact. After Wendy's began this unique social media approach in 2017, they experienced a 49.7% growth in profit that year.[3] Using the power of Internet memes, Wendy's was able to vitalize its brand even as an already successful company.

---

[1] https://twitter.com/Wendys/status/979022905039302661
[2] www.marketingdive.com/news/wendys-cmo-on-striking-the-right-amount-of-social-media-sass/538976/
[3] www.deputy.com/us/blog/how-wendys-used-social-to-profit-64m-in-a-year

© Griffin Kao, Jessica Hong, Michael Perusse and Weizhen Sheng 2020
G. Kao et al., *Turning Silicon into Gold*,
https://doi.org/10.1007/978-1-4842-5629-9_15

On the other side of the world, the Chinese multinational conglomerate Tencent was also able to achieve impressive growth with a mature product: its chat app WeChat. In China, WeChat is used as an entire ecosystem where one can talk with friends, play games, book appointments, and pay for anything from food to utility bills. As you can see, many of these actions involve payment, but at the time, not many people had entered their payment credentials into the app and were not as fully engaged with the WeChat ecosystem. Thus, in order to draw out the full potential of their app, Tencent needed to convince more of its users to provide payment credentials. In 2014, they achieved huge growth in usage by creating a feature that tripled the number of accounts with associated payment credentials on its debut.[4] This feature was a digital red envelope that you could use to send money, inspired by the physical red envelopes containing money that have been prevalent in Chinese culture since the Song Dynasty. These envelopes represent good luck and are exchanged between family and friends during social gatherings and holidays. By replicating this huge cultural practice in the digital world, WeChat was able to reach staggering amounts of usage. During Chinese New Year 2016, a whopping 32 billion envelopes were sent on the platform, a number which represents more than 20 times the size of China's population. There is consistently high usage even on non-holidays—every day, over 60 million people send and receive red envelopes.[5] By taking advantage of common cultural practices, WeChat was able to make an already popular platform even more widely used.

These two companies on opposite sides of the globe have seemingly little in common, yet they both employed shockingly similar growth strategies. From fast-food to social networking, both Wendy's and WeChat used cultural memes and viral tactics to create astonishingly fast growth.

The History of Memes

To understand the similarities in both stories, we must first introduce the concepts of a meme. Though the word "meme" is now largely used to describe media on the Internet, it was first coined by Richard Dawkins in the 1976 book *The Selfish Gene*. This book applied concepts from evolutionary biology to the study of information transfer. Much like how a gene is a unit of heredity that transfers from a parent to their offspring, Dawkins describes a meme as a unit of culture transferring from person to person.[6] Thus, the study of memetics was born.

---

[4] www.fastcompany.com/3065255/china-wechat-tencent-red-envelopes-and-social-money
[5] https://a16z.com/2016/07/24/money-as-message/
[6] www.smithsonianmag.com/arts-culture/what-defines-a-meme-1904778/

More specifically, a meme is defined as any cultural unit, such as an idea, belief pattern, or behavior, that spreads from person to person. In his book, Dawkins describes how memes live inside the mind of individuals and spread or "reproduce" by jumping into the mind of another person. Because ideas can combine with other ideas, memes are also able to fuse and evolve. Though the study of memetics has been criticized as being untested, unscientific, and incorrect (Dawkins himself eventually left the field of study), the term is still very widely used to describe quickly spreading pieces of information.

In the preceding examples, both Wendy's and WeChat leveraged cultural memes for growth because of the natural way that memes spread from person to person. Wendy's used literal Internet memes in its tweets and marketing campaigns to gain impressions and market their brand. WeChat exploited the already widespread behavior of giving red envelopes in Chinese culture. Both companies saw how cultural information spreads among people and wanted to harness that growth for their brands.

Understanding Virality

While memes describe *what* is spread, virality describes *why* things spread. Specifically, virality is defined as the tendency of consumers to spread information about products through their social networks, much like the way a virus spreads from one person to another. Viral marketing takes advantage of social networks to spread marketing messages, creating the potential for rapid multiplication. Just like a virus, a viral product can grow exponentially in the right environment.

In order to create viral growth for a product, you must build the viral loop into your product (Figure 15-1).

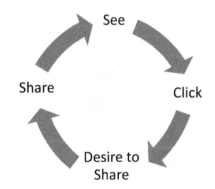

Figure 15-1. The Viral Loop

The viral loop describes a set of steps that a user goes through that will help naturally spread awareness of a product within their network. After a user sees a product, they will be compelled to click and share, eventually compelling more people to continue the cycle. Creating the foundation for this cycle in your product is crucial enabling continuous viral growth. Even with this foundation, for a piece of content to truly be viral, you must have a sufficiently high viral coefficient. The viral coefficient is defined as the average number of new users generated by an existing user. This can be calculated as

$$viral\ coefficient = \#\ shares\ per\ user \times \%conversion\ rate$$

In order to sustain viral growth, the viral coefficient must be greater than one. This means that, on average, each person that sees a piece of content must share it with more than one other person. If the coefficient is less than one, your initial set of users will share the product with a smaller group of new users. This second group of users will now share with a third even smaller group, and so on, until the groups of users that shares your product becomes smaller and smaller and eventually dwindles down to zero users leading to no more organic growth. When the viral coefficient is greater than one, new groups of users become larger and larger rather than decreasing to zero.

Virality is highly valuable as a business and marketing strategy because it provides organic growth with little cost. When done right, a viral product will grow on its own using the natural power of social networks. Virality does not use paid incentives to get people to share a product, as this is often unsustainable and will dry out once the money is gone. A viral product must survive on its own.

Another way to understand how virality works is to look at the marketing funnel (Figure 15-2).

Figure 15-2. Marketing Funnel

The marketing funnel describes the set of steps that users must go through in order to start using a product, beginning with the initial step of becoming aware of a product. In terms of the marketing funnel, virality helps build awareness and consideration for new users, often by facilitating advocacy from current users. In order to have conversion and loyalty, you must create a product your users legitimately value. Virality cannot be employed with bad products. In order to spread your product across society, you must have a product worth spreading.

As you can see, understanding both virality and memes shows us how to leverage social networks and create products that spread on their own. In the following paragraphs, we'll see exactly how WeChat and Wendy's were able to apply these concepts to reach millions of people.

Creating Viral Marketing: The Wendy's Twitter Account

While WeChat built a product, Wendy's built a marketing strategy. Though the two companies both depend on viral growth, there is a slight difference between creating viral products and using viral marketing. Marketing campaigns are made to promote products through different types of media. These campaigns can become viral when users are encouraged to share information about a product or company through their social networks. In order to create a marketing campaign that can become viral, you must address the following set of questions:

1. What will people be sharing?
2. Why will people share your content?
3. Where will your content be shared with new customers?

To get a deeper understanding of each question, we will use the Wendy's Twitter account as an example to break down each question.

What Will People Be Sharing?

Viral marketing must first have information that can go viral. This question is meant to pinpoint what type of information will be shared by customers. As with all marketing campaigns, the goal of the Wendy's campaign was to increase positive consumer awareness of the brand. In particular they decided to spread their playful brand personality through interactions with customers and other companies on Twitter, where they can share content demonstrating the brand's engaging personality. Through tweets, the company can answer user concerns, gather feedback, and promote new deals or products. Therefore, tweets offer the perfect bite-sized nuggets of information that can be shared with millions of people and go viral.

Why Will People Share Your Content?

In order for marketing content to go viral, there must be a reason people want to share it with others. This desire to share is a fundamental part of the viral loop and therefore an extremely important question to answer when designing viral marketing campaigns.

To get people to share their content, Wendy's focused on appealing to people's sense of humor and delight. Their tweets especially capitalize on the memetic properties of popular media trends and pop culture. This worked tremendously well—their tweet exchange with a teenager ended up earning him the most retweets of all time from 2017 to 2018.[7] Though there isn't an exact formula that can be replicated to determine what content will or will not go viral, the Wendy's Twitter account is highly responsive, and they have successfully developed a mascot that customers can truly engage with. In contrast to the stiff corporate personas of many other companies, Wendy's brand has successfully created a relatable and lighthearted personality without being cringe inducing. The people behind the tweets are clearly social media savvy and have a good understanding of what is trending on Twitter, without seeming behind the times or like they are trying too hard.

Using such a public platform definitely comes with risks, as consumers can be quick to judge when a tweet goes too far. As CMO Kurt Kane explains, "We want to be likable and sassy. We don't want to be seen as sarcastic and rude. But we walk a fine line. Sometimes we get it wrong in tone."[8] With such a public platform, Wendy's must be cautious of saying anything too polarizing that could reflect negatively on the company's brand. Though once in a while a tweet may fall flat, Wendy's is overwhelmingly able to generate a positive response to their tweets which leads to more people sharing their content.

Where Will Your Content Be Shared with New Customers?

After deciding on the type of content to create and figuring out how to motivate users to share your information, you need to figure out where this content will be shared with new customers. You want to make it easy for customers who see your content to quickly share it with others in their network.

---

[7] https://twitter.com/carterjwm/status/849813577770778624/photo/1
[8] www.forbes.com/sites/andriacheng/2018/10/08/wendys-twitter-roasts-have-become-the-envy-of-marketers-heres-how-it-does-it/#776abb8fea4c

Twitter has over 100 million users on the platform who could potentially be exposed to Wendy's content so it is an obvious choice when wanting a platform to engage with new potential customers.[9] Furthermore, users can retweet and share content to their own followers, thus providing a perfect platform for virality.

Answering the three questions will help you address the different components of the viral loop. In order to create a marketing campaign that will naturally spread among customers, you must decide what kind of information to share, understand why your customers will want to share it, and ensure that your customers are able to easily share it with other people who have not yet been exposed to it. If these questions are successfully addressed, your marketing campaign has the components to go viral, provided that users respond positively.

Building Virality Into Products: WeChat's Red Envelope Feature

While Wendy's shows us how to build viral marketing campaigns, we can look to WeChat to better understand how to build virality into a product itself. To create the viral loop in a marketing campaign, we discussed three questions that needed to be addressed. In order to create the viral loop in a product, we must add a fourth new question that needs to be answered. The new set of questions is

1. What will people be sharing?
2. Why will people share your product?
3. Where will your product be shared with nonusers?
4. Why will exposed nonusers use your product?

Let's break each of these down further and see how WeChat was able to address each question.

What Will People Be Sharing?

To create a viral product, you must have something that can be shared in the first place. For WeChat, this is the red envelope feature. This is where the cultural meme is highly valuable, as they reference a unit of culture that is

---

[9] www.statista.com/statistics/282087/number-of-monthly-active-twitter-users/

already prone to being spread. The digital red envelopes were built in reference to the long-standing cultural tradition of sharing physical red envelopes, thus making a digital red envelope the perfect choice for a viral product.

Why Will People Share Your Product?

This question is truly at the heart of virality in product design. Broadly, there are two main reasons that people will share a product. In the first case, users simply gain value from sharing the results of the product. A prime example is sharing Instagram posts on Facebook to easily gain more exposure. This is known as **distribution virality**. In the second case, existing users gain value when more of their network uses the product, thus incentivizing them to get others to use it as well. This is known as **pull virality**, as nonusers are pulled by current users into using your product. For example, messaging platforms are only useful when you have people to message, and work collaboration tools only work when your coworkers also use them. When building virality into a product, you can choose to employ either or both of these approaches.

Oftentimes, there is also an innately emotional reason that people share products. This requires you to deeply understand the psychological motives behind why a user would share your product, whether because it helps address an innate need to belong or because it develops self-esteem. Using an emotional driver for your product can help you appeal to users on a more fundamental level.

WeChat's red envelopes rely on distribution virality, as users inherently gain value by being able to send envelopes to their friends. As a product that digitizes a cultural tradition, there are many different emotional levers it uses to incentivize sharing as well, including demonstrating generosity, communicating affection, and connecting with loved ones.

Where Will Your Product Be Shared with Nonusers?

Once we have determined why people will share your product, we need to figure out how to use these behaviors to spread awareness of your product. This means the product must be shared with nonusers, as sharing among only current users leads to no growth. Therefore, you need to identify easy ways for the product to be shared with nonusers. Oftentimes, users will be sharing products through other forms of social media, which is why it is important to integrate with channels that people already use. The experience of sharing your product on other forms of social media should be as frictionless as possible.

As an already mature product, WeChat was able to use its own platform to spread the usage of red envelopes. Because they were designed to spread adoption of payment credentials on the platform itself, WeChat only had to ensure that the process of sharing on the platform was as easy as possible.

Why Will Exposed Nonusers Use Your Product?

The final question to close the viral loop is to address conversion. The recipients of your product may often ignore what is shared, especially in the distribution model. Therefore, you need to immediately communicate value of your product when it's shared and make sure you are addressing true user needs. Emotional drivers can also be useful at this stage, as seeing how others benefit can incentivize your own usage. For example, in WeChat's case, more and more people decided to enter their payment credentials to use the product because of the many emotional and cultural factors.

Overall, these four questions provide a framework that will help you build virality into your product. WeChat was able to address all four questions with the red envelopes and therefore had the necessary elements to make their product go viral.

Both Wendy's and WeChat demonstrate the power of cultural memes in viral marketing campaigns. A cultural meme provides the perfect fodder for virality because of its innate tendency to be shared. By addressing the sets of questions explained above, you can build a product or marketing campaign that can promote the viral loop and lead to explosive organic growth.

Takeaways: Overall, both Wendy's and WeChat were able to build campaigns and products that take advantage of the properties of cultural memes to go viral. In order to build viral products, you must address the questions of the viral loop to ensure people become aware, share, and repeat.

Questions

1. How do you ensure virality creates sustainable growth for a product, rather than just creating a passing fad? What retention mechanisms do you need in place?

2. Virality as a concept can clearly be applied to social media. In what other industries and product areas can virality be used?

Failing

Un-assistive Voice Assistants

Lessons Learned in History

Amazon introduced eight new Echo devices in 2019, including Echo Buds (wireless earbuds), Echo Frames (smart glasses), and the Echo Loop (a smart ring).[1] All variations of Echo offer hands-free access to Alexa, Amazon's virtual assistant who can assist users with anything from playing music to controlling your smart home to buying other products from Amazon. In the same year, Amazon also launched the "Voice Interoperability Initiative" to integrate numerous virtual assistants in one single device and provide consumers easy access to their multitude of assistants (though notably missing some of its top competitors, such as Google's Assistant, Apple's Siri, and Samsung's Bixby).[2] Many other tech giants, including Google, Apple, Samsung, and more, are also frequently making the headlines with new assistant technologies.

---

[1] https://press.aboutamazon.com/news-releases/news-release-details/amazon-introduces-8-new-echo-devices
[2] www.cnet.com/news/amazons-new-initiative-wants-to-make-it-easier-to-use-multiple-voice-assistants/

© Griffin Kao, Jessica Hong, Michael Perusse and Weizhen Sheng 2020
G. Kao et al., *Turning Silicon into Gold*,
https://doi.org/10.1007/978-1-4842-5629-9_16

It seems like the age of virtual assistants. Virtual assistants (also commonly referred to as voice assistants or intelligent personal assistants) have become one of the hottest tech trends of today.

The Beginnings of Smart Assistants

Though a seemingly recent development, virtual assistants have been evolving for the past century. In fact, the concept of voice assistants has been around since 1911, when the Radio Rex was released.[3] The toy was an early attempt at speech recognition consumer electronics, where a toy dog would respond to its owners' calls of "Rex!," delighting children and adults alike.

Many decades later, the technology showed substantial progress. The 1961 IBM Showbox was a computer that recognized 16 spoken words, including the digits 0 through 9 and mathematical operations like "plus" and "minus" and was able to perform simple arithmetic problems.[4] The true foundation for modern virtual assistants was laid out in 1994 with the launch of the first smartphone, known as the "IBM Simon," which came with a personal digital assistant (PDA).[5] PDAs were the ancestor of today's virtual assistants, providing users information management capabilities such as calendars and address books. Since then, assistants have grown and evolved beyond these basic capabilities.

Apple's Siri was officially announced to the world in 2011, but Siri first debuted in 2010 as an independently developed iPhone assistant app created by a 24-person start-up.[6] The trio that cofounded Siri envisioned it to be a platform that would allow users to converse with the Internet; its initial capabilities included connecting with 42 web services (i.e., Yelp, StubHub) and providing users the ability to use the functionalities of these web services (i.e., reserving a table, buying tickets) without needing to open the app.[7] The app was meant to define the third generation of the Web, replacing the "search" functionality with the "do" functionality. Instead of users searching the Web for restaurant recommendations and reservations, they could use Siri to directly make the reservation.

Although Verizon sought to make Siri a default app on all Android phones, Apple acquired Siri and made the technology exclusive to Apple devices.[8] From there, Siri became the first virtual assistant available to the masses and

---

[3] www.cnet.com/news/google-finding-its-voice/
[4] www.ibm.com/ibm/history/exhibits/specialprod1/specialprod1_7.html
[5] www.theinquirer.net/inquirer/news/2360484/ibms-simon-smartphone-is-20-years-old
[6] www.huffpost.com/entry/siri-do-engine-apple-iphone_n_2499165
[7] Ibid.
[8] Ibid.

sparked today's voice assistant revolution. Despite the drastic improvements made to the smart assistant in the past century, it has been a bumpy journey to develop the perfect assistant—and it still appears to be far from over.

Lessons Learned from Clippy

This brief history of smart assistants skips over an infamous example that many of us likely don't associate with modern voice assistants. Long before Siri ambitiously entered the stage, Microsoft created a widely known and widely hated virtual assistant: Clippit. Known colloquially as Clippy, the virtual paperclip was introduced in 1996 and welcomed you as you opened Microsoft Word.[9] Clippy was the reincarnation of Microsoft Bob, an assistant launched in 1995 that was designed to make software friendlier. Microsoft Bob's library of animated assistants, including a dog named Rover and a rat named Scuzz, was unfortunately a huge miss because of its perceived childishness and was quickly abandoned by Microsoft.[10] Clippy was meant to achieve the same vision that Bob once had, transforming feature-heavy programs into something much lighter and easier for users.

Microsoft used Stanford studies to develop their assistant, which found that humans had more positive computer experiences if the computers responded as if they were human.[11] Despite their best efforts, Clippy too was a failure and was met with animosity from users. Examining more closely why this happened, we find that Clippy faced three core problems:[12]

1. **Functionality:** One of Clippy's biggest issues was that he simply wasn't useful. Although somewhat useful for a first-time user, Clippy became frustrating for most other users with his (unhelpful) attempts to write a letter or interruptions via knocking on the screen.[13] Despite being a solution to unfriendly computer software, Clippy was sadly not a very useful one.

---

[9] http://mentalfloss.com/article/504767/tragic-life-clippy-worlds-most-hated-virtual-assistant
[10] Ibid.
[11] www.artsy.net/article/artsy-editorial-life-death-microsoft-clippy-paper-clip-loved-hate
[12] https://medium.com/@uneeq/what-went-wrong-with-clippy-the-virtual-assistant-pioneer-people-loved-to-hate-5a8cac684778
[13] Ibid.

2. **Design choices:** Clippy was also not designed with the quality of the user experience in mind. When developing an assistant, it's vital to consider how users feel when interacting with it. At the time Clippy was launched, Microsoft had a male-heavy leadership that ignored focus group testing results indicating that women found the design of Clippy to be somewhat disturbing or uncomfortable.[14] The result was an assistant that was not only unhelpful but also created a distressing user experience.

3. **Personalization and humanization:** Clippy lacked humanity and wasn't personalized for users at all. Clippy couldn't learn a user's name or preferences, no matter how many times they interacted.[15] As an assistant meant to respond like a human would, this lack of humanity was definitely damaging.

In combination, these three issues prevented Clippy from becoming a powerful and helpful assistant that its users could rely on. Fast-forwarding to the present day, the question remains if companies can overcome these problems to create the truly reliable assistant that we all envision.

Assistants in the Age of Ambient Computing

We are currently in an era of ambient computing, a concept where humans can use technology and devices anytime, anywhere. Technology is constantly around us and constantly responding to our needs, creating an environment where computers and electronics are always at the tips of our fingers. With this current trend, virtual assistants play a key role. They offer a way of using technology hands-free, without needing to see or touch devices to use them. In essence, they share a common goal with our old friend Clippy: to make technology and software friendlier and easier to use. Numerous tech behemoths and start-ups have jumped on the bandwagon to develop these assistants and provide another way for users to interact seamlessly with technology.

As a result, nearly 34% of the US population used voice assistants monthly as of 2019, with that number projected to continue growing in the next few years.[16] Unfortunately, the story reveals a less optimistic perspective once we

---

[14] https://medium.com/@uneeq/what-went-wrong-with-clippy-the-virtual-assistant-pioneer-people-loved-to-hate-5a8cac684778
[15] Ibid.
[16] www.searchenginejournal.com/33-of-people-are-now-using-voice-assistants-regularly/321413/#close

look more closely. Although voice assistant use grew 9.5% in 2019, that number seems rather feeble in comparison to growths of 27% and 24% in 2018 and 2017, respectively.[17] Voice assistants may still be growing, but the numbers reflect that the technology still needs some more work before it can truly meet and satisfy user needs. In fact, today's smart assistants face variations of many of the problems that plagued Clippy as well, all of which need to be addressed before these embedded technologies can truly become the all-helpful assistants that the tech industry envisions.

1. **Functionality:** Voice assistants now offer far more impressive abilities than Clippy once did. Unfortunately, this is a double-edged sword and has forced consumer expectations to grow beyond what these assistants are actually capable of. Voice assistants use natural language processing (NLP), a form of machine learning, to parse and understand spoken voice commands. This means they're using machine learning models trained on specific skills and abilities and tend to perform these particular skills extremely well. As of 2018, Amazon's Echo had over 30,000 skills available to users.[18] However, as with all deep learning algorithms, voice assistants can only excel in the specific domain they were trained for. As soon as users start issuing new, foreign voice commands, these assistants have no way to know how to respond.

    Unlike visual displays and graphical user interfaces (GUI), voice user interfaces (VUI) don't provide clear boundaries for what they are and aren't capable of accomplishing.[19] Visual screens make it clear what applications and buttons can be tapped on and interacted with; on the other hand, voice interfaces have no way of doing so. In fact, 49% of users have said they don't know where to begin accomplishing voice tasks.[20] Users simply don't know what skills their assistant is limited to and, as a result, often ask it to do tasks it cannot do. This has spawned hashtags like #AlexaFails to reflect user frustration with assistants that aren't as powerful as they initially seem. Instead, users can only trust their assistants to reliably perform simple tasks. In fact, studies have found that the majority of tasks being done on voice assistants are asking

---

[17] Ibid.
[18] https://bdtechtalks.com/2018/09/03/challenges-of-smart-speakers-ai-assistants/
[19] Ibid.
[20] https://theblog.adobe.com/voice-assistant-statistics-trends-2019/

basic questions, searching/requesting information, and/or listening to music—all tasks that can be easily done without voice as well.[21] Though functional, voice assistants don't provide valuable function—the tasks that they are able to perform aren't actually made that much easier via voice interface.

2. **Design choices:** Although today's voice assistants have emphasized good user experiences, they are still facing some design problems. In particular, many of today's prevalent voice assistants are being critiqued for having female voices.[22] This pattern subconsciously reinforces certain gender stereotypes, and although the Google Assistant and Siri now offer the option to switch to male voices, the default design choice reveals some of the same sexist biases that Clippy also demonstrated.

3. **Personalization and humanization:** On the flip side of the lack of sufficient personalization, voice assistants today are commonly criticized for invading users' privacy in their attempt to mine data and become more personalized. The issue of voice assistant privacy warrants its own discussion, with companies like Google and Amazon reported to have been analyzing (anonymized) voice assistant recordings in 2019.[23]

The Future of Voice Assistants

Despite the shortcomings in today's technology, given the current popularity of voice assistants in the market, it's clear that they are here to stay, at least in the short term. VUIs are a naturally intuitive way for humans to interact with technology, given that voice is something we use in our daily lives. Translating this to consumer technology is proving to be difficult, but given that the technology is still in its relative infancy, there is still room for a breakthrough to happen soon. These assistants' capabilities are growing by the day, and though we have smart glasses and smart rings that some may call gimmicky today, perhaps we'll have truly trustworthy digital assistants in the very near future.

---

[21] Ibid.

[22] https://medium.com/pcmag-access/the-real-reason-voice-assistants-are-female-and-why-it-matters-e99c67b93bde

[23] www.theguardian.com/technology/2019/oct/09/alexa-are-you-invading-my-privacy-the-dark-side-of-our-voice-assistants

Takeaways: Virtual assistants are a huge technological trend but still often fail to meet user needs and expectations by solving real pain points. Although there have been several milestones in the history of digital assistants, many of the problems plaguing pioneering virtual assistants still exist today in some form. Companies working on voice assistants need to keep in mind that they need continuous innovation to ensure voice assistants will provide real value for users.

Questions

1. Are there particular situations where you wish you could use technology hands-free? Would you prefer to use technology via voice commands in those cases?

2. How do you feel about interacting with computers in the same way that you interact with other humans? How much would you want to consider your virtual assistant as a friend vs. an emotionless helper?

Too Much of a Good Thing

Why Are There So Many Social Messaging Apps?

"Text me."

The meaning of this simple, commonplace phrase is subtly shifting as the next generation of Gen Z'ers move toward using social messaging apps over traditional SMS (Short Message Service) texting. With 52% of teens now using these messaging apps for over 3 hours per day,[1] the action of "texting" or "messaging" someone is becoming overloaded and could refer to a dozen different messaging apps. How many social messaging apps do *you* have installed on your phone?

---

[1] *www.thinkwithgoogle.com/interactive-report/gen-z-a-look-inside-its-mobile-first-mindset/*

© Griffin Kao, Jessica Hong, Michael Perusse and Weizhen Sheng 2020
G. Kao et al., *Turning Silicon into Gold*,
https://doi.org/10.1007/978-1-4842-5629-9_17

The Messaging Market Today

Chances are, you have a least a handful of the most popular messenger apps (statistics as of July 2019)[2]:

- WhatsApp (1.6 billion monthly active users, or MAUs)—a cross-platform messaging and voice over IP (VoIP) app owned by Facebook; popular in multiple countries (the United Kingdom, Spain, India, etc.)[3]

- Facebook Messenger (1.3 billion MAU)—a messaging app spun off from the core Facebook app; popular primarily in North America[4]

- WeChat (1.1 billion MAU)—a messaging and social media app developed by Tencent; China's most popular messaging app[5]

- QQ Mobile (800 million MAU)—another huge social media platform developed by Tencent; China's first popular social media app[6]

- Snapchat (290 million MAU)—a photo messaging app; popular primarily in North America and Europe[7]

- Viber (260 million MAU)—a messaging app developed in Tel Aviv and owned by Japanese e-commerce conglomerate Rakuten; popular in Eastern Europe and Russia[8]

- Discord (250 million MAU)—a messaging app specifically designed for and popular in the gaming community[9]

- Telegram (200 million MAU)—a messaging app founded by a Russian entrepreneur with a focus on privacy; the most popular messaging app in Ethiopia, Iran, and Uzbekistan[10]

---

[2] www.statista.com/statistics/258749/most-popular-global-mobile-messenger-apps/
[3] www.messengerpeople.com/global-messenger-usage-statistics/
[4] www.socialmediatoday.com/news/facebook-messenger-by-the-numbers-2019-infographic/553809/
[5] www.cnbc.com/2019/02/04/what-is-wechat-china-biggest-messaging-app.html
[6] www.clickz.com/qq-the-biggest-digital-platform-youve-never-heard-of/113476/
[7] https://mashable.com/article/snap-q2-2019-earnings/
[8] www.messengerpeople.com/messaging-apps-brands-viber-messenger/
[9] www.businessinsider.com/how-to-use-discord-the-messaging-app-for-gamers-2018-5
[10] www.businessofapps.com/data/telegram-statistics/

This list is far from comprehensive and doesn't include other common messaging apps like the default texting app on mobile phones ("SMS" texts), iMessage (Apple's messaging service), Skype (an app specializing in video chatting), GroupMe (a mobile group messaging app owned by Microsoft), and more. Depending on your view, social media apps like Twitter, Instagram, and even LinkedIn (a "professional" social media platform) could also be considered messaging apps. We could also include workplace messaging apps, such as Slack or Stride, or other niche messaging apps like Twitch (a video live-streaming service with a private messaging feature).

It doesn't take a hard look to see the abundance of messaging apps that exists in today's online community. Even by looking at a single college campus, we can see that different groups and clubs use a huge variety of messaging apps, with some preferring to use iMessage, some opting for Facebook Messenger, and still others choosing GroupMe. But given our social tendencies, most people are part of more than just one single group—instead, our complex social networks require that we juggle an equally complex network of social messaging apps and platforms, with 41% of American adults using two or more messaging services in 2018.[11]

Social Messaging vs. Social Media Platforms

Interestingly, messaging apps are dominating social media platforms. Although social media networks have been getting most of the attention in the media, the 3.4 billion users of the four largest social networks (Facebook, Twitter, Instagram, LinkedIn) in 2018 actually totals less than the 4.1 billion users of the four largest mobile messaging apps (WhatsApp, Facebook Messenger, WeChat, Viber) in the same year.[12] Statistics like the number of messages sent are equally eye-opening; in 2018, 72 trillion messages were sent across these 4 platforms, a number that greatly exceeds even the 1.6 trillion searches on Google in the same time period.[13] The world is communicating through these relatively new platforms.

While the frequency at which we send electronic messages may be unsurprising, the influence that social messaging apps wield even on the platforms that host them should be. It seems like any online platform now requires a chat system as a bare minimum requirement, with web sites like LinkedIn, Instagram,

---

[11] www.statista.com/statistics/946764/number-messaging-services-used-adults-past-month-usa/
[12] www.adweek.com/digital/heres-how-messaging-is-positioned-to-dominate-in-2019/
[13] Ibid.

and Tumblr incorporating social message functionalities in recent years.[14,15,16] Where is this obsession coming from?

A huge reason may be the growing trend of consumers valuing private social interactions more so than public social media posts. Controversy has surrounded these tech platforms due to privacy issues (Facebook's numerous scandals in recent years), resulting in people seeking social activity in more one-on-one, personalized settings—in other words, messaging apps. This transition to private social activity explains the staggering number of users that we see using messaging apps, beyond those of social media.

This Internet phenomenon signals a shift among businesses as well, most notably for marketing teams. As companies find their online personas and define their marketing campaigns, they must now also consider directing that strategy into more private messaging channels. We already see a rise in technologies such as chatbots that are being used to redefine customer engagement via real-time interactions,[17] all from the customer's convenience at home. Then this begs the question, where is the line? Will companies drive away loyal customers by infringing on a space once valued for being free from the ads that plague the rest of the world?

The Fragmented Nature of Messaging

But perhaps the more interesting (and easy-to-answer) question is not how companies should tailor their marketing strategies around this cultural change but, rather, understanding why and how so many social messaging apps can coexist. Since the early days of the 2000s when SMS (the traditional messaging app) was conceived, messaging has evolved and transformed into the complex network of apps that exists today. As the world moved online, countless companies joined in the fun and contributed their own version of a messaging app, hoping to capture consumers' desire to stay connected with the people in their lives via a personalized and private channel.

These consumers exist all across the globe and come from all walks of life. Importantly, they also live in different countries, have different cultures, and have different social norms. The success of a messaging app in a given locale is highly dependent on the app being localized to the consumers there.

---

[14] www.socialmediatoday.com/news/linkedin-rolls-out-new-messaging-tools-to-enhance-on-platform-connection/528235/
[15] www.socialmediatoday.com/news/instagram-is-developing-a-separate-messaging-app-called-threads/561797/
[16] https://techcrunch.com/2015/11/10/tumblr-rolls-out-instant-messaging-on-both-web-and-mobile/
[17] www.adweek.com/digital/heres-how-messaging-is-positioned-to-dominate-in-2019/

A prime example is WeChat's unparalleled dominance in China. What started as a simple messaging app has now evolved into a ubiquitous "super app" with countless features baked in, from preordering food at restaurants to booking taxis to making payments. Tencent's app became wildly successful in China because it was designed uniquely for Chinese consumers—Tencent took into account the Chinese cultural attitude toward money and turned WeChat into a mobile wallet and also capitalized on the close relationship between Chinese corporations and the Chinese government, allowing the government to fully support and nurture WeChat into an all-encompassing lifestyle app.[18] These factors, which drove WeChat's domestic success, are completely non-applicable in the global market.

The LINE app is another great example. Naver Corp's messaging app is on over 90% of all smartphones in Japan because of how it fully captures Japanese user needs.[19] LINE is a direct social network, unlike group social networks such as Facebook—this means that content cannot be spread widely and instead is shared only through private channels. This design choice clearly reflects a Japanese culture that highly values privacy. Furthermore, stickers are an integral feature of LINE and form a massive marketplace on the platform. This is relevant in Japan, where computer users have been using emojis since the 1980s. LINE's stickers are today's modern incarnation of emojis, with a recognizable cast of characters and personalities that embody the Japanese concept of *kawaii* (the Japanese obsession with all things cute).[20] Again, these distinct qualities of LINE make it uniquely suitable for the Japanese market.

WhatsApp is an interesting example of a product having great timing. It's a huge hit in emerging markets like India, where users were coming online for the very first time via their phones.[21] Here, there weren't any preexisting social networks like Facebook or Twitter. Instead, users picked up WhatsApp, a clean, ad-free messaging platform. WhatsApp also gained traction in Europe, where there weren't any unlimited text plans like in the United States. WhatsApp offered a cheaper alternative for texting, while still maintaining the same feel.[22]

These messaging apps which found their roots in niche geographic pockets have also found ways to branch out to other locales through immigrants, according to the app-tracking company Onavo.[23] Immigrants continue using

---

[18] https://blog.prototypr.io/why-chinas-super-apps-will-never-succeed-in-the-us-64c686c8c5d6
[19] www.humblebunny.com/line-japans-favorite-mobile-messenger-app/
[20] www.fastcompany.com/3041578/how-japans-line-app-became-a-culture-changing-revenue-generat
[21] www.wired.com/2014/02/whatsapp-rules-rest-world/
[22] www.wired.com/2014/02/whatsapp-rules-rest-world/
[23] https://techcrunch.com/2013/06/13/messaging-apps/

messaging apps from their home countries to stay in touch with friends and family still living there; as a result, they're bringing these apps to new countries, often where other messaging apps also already exist. In today's increasingly globalized world, the messaging world continues to grow more and more fragmented.

The Future of Messaging

Despite the localization of these messaging apps, there's no denying that many of them now offer the same core functionalities. There's no key difference between WhatsApp and Facebook Messenger, and some analysts expect some cannibalization to occur among social messaging apps until the average user uses fewer than two active apps.[24] Whether this means Facebook manages to buy and expand its social messaging portfolio (it currently has the impressive lineup of Facebook Messenger, WhatsApp, and Instagram) or some competitors begin leaving the scene, the market of social messaging apps is likely reaching saturation. With users owning such varied messaging platforms, it is becoming overwhelming to keep track of which social interactions happen in which app. Given the difficult and fragmented nature of messaging though, it's a distant reality that any messaging monopoly will emerge among even tech giants anytime soon.[25]

Companies are noticing and responding to consumer trends of valuing private social interactions, but there comes a point where we must ask ourselves—how much is too much? How many social messaging apps must we constantly switch between just to navigate our social networks? In some ways, texting has regressed since its humble days of SMS, where things were simple and "text me" had only one possible meaning.

Takeaways: There are numerous messaging platforms with subtle differences. These products should reflect trends in consumer preferences, as illustrated by the shift from social networks to social messaging. At the same time, companies need to fully understand their users to capture the market, as illustrated by the various localizations of messaging apps in different countries.

---

[24] https://ark-invest.com/research/social-messaging-apps
[25] www.vice.com/en_us/article/zma735/why-do-we-need-so-many-different-messaging-apps

Questions

1. What convinces you to join a new social platform/messaging app? What makes you abandon a social platform/messaging app?

2. How niche should messaging apps be? Do you want different communication methods between friends and coworkers? Family and friends? Close friends and acquaintances?

3. Imagine you're in charge of marketing at a large firm—would you utilize private messaging channels to advertise? If so, how would you do it?

4. Can you envision a unified messaging app used globally by all users? What kind of features would it need? Is this kind of app even possible, given the world's numerous different cultures and societies?

print("No.")

Why Tech Employees Are Able to Dictate Company Direction

"Don't be evil."

This has been Google's unofficial motto since early 2000, and despite its playful simplicity, the company and others like it have found themselves stewing in controversy.[1] But outside of traditional **techlash** (a portmanteau of "tech backlash") surrounding the growth and control that these large tech companies can exert of their users—the world—a new breed of *internal* controversies between employees and their companies has emerged.

When companies like Microsoft employ over 140,000 people and operate in over 100 countries, it's not surprising that disagreement among the ranks is an issue. But when internal issues are founded on questionable ethical decisions from the broader executive level, the outer world takes notice, and so do investors. In this chapter, we will look at several instances of internal controversies within Amazon, Microsoft, and Google, explaining the events that led to internal friction and the responses that employees took to affect change. Then, we will consider the types of environments that large tech companies brew as employers, assessing what circumstances might be most responsible for determining if heightened employee activism will flop or change the company's direction.

---

[1] https://abc.xyz/investor/other/google-code-of-conduct/

© Griffin Kao, Jessica Hong, Michael Perusse and Weizhen Sheng 2020
G. Kao et al., *Turning Silicon into Gold*,
https://doi.org/10.1007/978-1-4842-5629-9_18

Amazon

It's clear that Amazon has had a global impact on how people experience e-commerce, but employees feared that the impact was having a global impact on the environment as well. On Friday, September 20, 2019, thousands of Amazon employees across 14 countries walked out to protest Amazon's impact and inaction on climate change.[2]

The protestors had three main demands:[3]

1. That Amazon would commit to zero carbon emissions by 2030

2. That Amazon would commit to zero funding for politicians who denied climate change

3. That Amazon would commit zero work on AWS (their cloud Amazon Web Services) for oil and energy companies contributing to climate change

These demands were made known prior to the walkout, where activists circulated internal materials that made their way up the ranks to Jeff Bezos himself.

Bezos responded to the demands on the Thursday prior to the walkout, in Washington, DC, where he pledged that Amazon would abide by the Paris Climate Agreement—an international pledge for countries to go carbon neutral by 2040.[4] **Carbon neutral** means that whatever carbon emission is made would be repaid, in the form of carbon savings somewhere else in the world. For Amazon, this means that whatever they expend on carbon emissions for things such as shipping, they would remove carbon elsewhere, by planting many trees for instance.

Bezos made no promises for how Amazon would change its political donations, as he was skeptical that any money was making it to antagonists or skeptics of climate activism. He also made no promises for changing how the company offers AWS to oil and gas companies, saying that "[asking] oil and energy companies to do this transition with bad tools is not a good idea and we won't do that."[5]

---

[2] www.cnn.com/2019/09/20/tech/amazon-climate-strike-global-tech/index.html
[3] www.vox.com/recode/2019/9/20/20874497/amazon-climate-change-walkout-google-microsoft-strike-tech-activism
[4] www.nytimes.com/2019/09/19/technology/amazon-carbon-neutral.html
[5] www.nytimes.com/2019/09/19/technology/amazon-carbon-neutral.html

Overall, the walkout had been a success, and workers at Microsoft and Google followed suit with similar protest strategies.[6] Getting Amazon to think critically about its involvement with climate change in front of the public eye offered the right environment of internal and external pressure to affect change from the Bezos level. Notably, this walkout occurred amidst the company culture of Amazon that is described firsthand as a challenging place to make objections, as the company prides itself in solidarity and secrecy.[7] The fact that these tech workers chose to even stage the walkout might have been a consequence of them following in the footsteps of trailblazing activists at rival tech companies.

Microsoft

While Amazon's September walkout focused on environmental issues, Microsoft employees in February 2019 objected to their company's $480 million contract from the US Army to build the Integrated Visual Augmentation System (IVAS). 100,000 headsets (at a whopping price of about $4,800 each) would be built to aid soldiers in combat and training, using modified versions of the HoloLens.

For background, the HoloLens is Microsoft's augmented reality (AR) headset. The device looks like a combination between a visor and glasses and users wearing it are able to see computer graphics elements embedded into their field of view. An artist might use the HoloLens to paint objects in 3D and walk around to visualize his or her creation. A director might use the HoloLens to see all monitors that his or her cameras are seeing, as well as the day's shoot schedule. But for this instance, Microsoft would be building out the HoloLens' features to aid soldiers in training and combat, offering a similar viewing experience to what *Call of Duty* games show players on the screen while in game—such as a map, if a person is an allied soldier, what your gun is aiming at, etc.[8]

Microsoft employees responded to the contract by drafting a public letter to Satya Nadella and Brad Smith—their CEO and President, respectively—titled "HoloLens for Good, Not War."[9]

"As employees and shareholders we do not want to become war profiteers," the letter reads. "We did not sign up to develop weapons, and we demand a say in how our work is used." This letter emphasizes a key component in tech employee rationale behind their protests; there is a mismatch between the

---

[6] https://fortune.com/2019/09/16/global-climate-strike-protest-google-amazon-microsoft-walkout/
[7] www.nytimes.com/2015/08/16/technology/inside-amazon-wrestling-big-ideas-in-a-bruising-workplace.html
[8] www.theverge.com/2019/4/6/18298335/microsoft-hololens-us-military-version
[9] https://twitter.com/MsWorkers4/status/1099066343523930112

promise of purposeful work that helps others and occasional controversial projects that attempt to justify that sentiment. Microsoft's mission is to "empower every person and organization on the planet to achieve more," but when "empower" and "achieve more" mean enhanced lethality and "every person and organization" are soldiers in the US Military, the dissidence is jarring.

But despite these protests, Microsoft leadership has stood its ground, and the development of the IVAS headset has reached the testing phase. Microsoft CEO Satya Nadella replied to the letter in an interview, stating that "we have made a principled decision that we're not going to withhold technology from institutions that we have elected in democracies to protect the freedoms we enjoy. We were very transparent about that decision and we'll continue to have that dialogue."[10]

This follows a track record of Microsoft standing its ground in providing software and hardware for the US government and military. Nadella made similar justification for the use of providing ICE (the US Immigration and Customs Enforcement) with cloud services and for signing the $10 billion JEDI (Joint Enterprise Defense Infrastructure cloud) defense contract with the Pentagon, despite similar employee outcries.

Microsoft has clearly chosen to take a rigid stance against employee concerns. For the HoloLens contract, this was a calculated risk; military money could propel the development of the HoloLens to a point that makes the product better and cheaper for consumers, but how many times can a company frustrate its employees before losing morale and popularity among potential workers such as college new graduates and student interns?

Google

The search giant has experienced perhaps the greatest number and variety of internal controversies. Among the first was a November 2018 walkout where over 20,000 employees walked out to protest how Google handles cases of sexual harassment.[11] Employees also walked out at the time Amazon employees walked out in September 2019 to protest their respective companies' impact on climate change (the day before this, Google CEO Sundar Pichai strategically announced their "biggest renewable energy purchase ever").[12] Google also dropped out of a bidding war with Amazon and Microsoft for the $10 billion JEDI defense contract with the Pentagon following employee

---

[10] www.theverge.com/2019/2/25/18240300/microsoft-ceo-defends-pentagon-contract-ar-headsets-employee-outcry

[11] www.npr.org/2018/11/01/662851489/google-employees-plan-global-walkout-to-protest-companys-treatment-of-women

[12] www.blog.google/outreach-initiatives/sustainability/our-biggest-renewable-energy-purchase-ever/

backlash, stating that the contract "conflicted with its corporate values."[13] Perhaps most controversial to outside eyes was Google's August 2018 consideration of developing "Dragonfly," the project codename for an Internet search that would be designed to be compatible with China's censorship laws. Pichai stated that the plans were purely "exploratory," and internal and external backlash ensured that such efforts remained exploratory.

Amidst these moral crises, a particularly surprising corporate decision provoked intense internal backlash: Google's instantiation of an AI & Ethics Advisory Board. The panel would provide an external report on whether or not Google was using AI responsibly, using expert advice to guide their development of AI technologies.

Formally dubbed the "Advanced Technology External Advisory Council," it fielded several academics from philosophy, robotics, to CS/AI departments at esteemed universities. It also had Kay James Cole, the president of the Heritage Foundation. In an open letter titled "Googlers Against Transphobia and Hate" that thousands of Google employees signed, employees called for the removal of Kay James Cole for being "vocally anti-trans, anti-LGBTQ, and anti-immigrant." Other members of the panel resigned following the outcry, including Alessandro Acquisti, a professor at Carnegie Mellon who studies technology ethics. In a tweet, Acquisti announced his resignation from the panel: "While I'm devoted to research grappling with key ethical issues of fairness, right & inclusion in AI, I don't believe this is the right forum for me to engage in this important work."[14] Only one week after its conception, Google decided to dissolve the panel.

Even amidst an attempt to preempt controversy surrounding the use cases of AI, Google found itself facing internal controversy surrounding the people it chose. This highlights the importance of getting every detail "right" in making pushes for positive change, where the optics of every decision are up for employee, shareholder, and user scrutiny.

Big Tech = Big Controversy?

Technology companies now more than ever are under an ethical microscope as they attempt to innovate, optimize, and expand their reach in competitive markets. Corporate leadership is being held accountable by the thousands of employees that represent and build out their company brands across the Web and into billions of people's lives. Whether or not the tech industry faces continued controversy will be determined by how these companies make

---

[13] www.bloomberg.com/news/articles/2018-10-08/google-drops-out-of-pentagon-s-10-billion-cloud-competition
[14] https://twitter.com/ssnstudy/status/1112099054551515138

trade-offs between transparency of decision-making and the speed of their growth. Companies are incentivized to move fast to innovate, yet hasty work is much more likely to result in such controversies that do not have approval internally or externally. How employees continue to respond to internal decisions and dynamics will be an early predictor to the technological and ethical directions the industry is taking.

Takeaways: Tech companies have found themselves facing "techlash" due to social, ethical, and environmental controversies over the past decade. Employees are often the first line of protest against company-wide contracts, operations decisions, and ethical alignments.

Questions

1. As a CEO, what metrics would you prioritize in using to determine whether or not a potentially risky opportunity would be worthwhile?

2. For companies facing recent controversy, what are actionable steps toward improving public image toward users, investors, and potential employees?

The Rise and Fall of Virtual Reality

What It Takes to Create Technological Disruption

In *Pygmalion's Spectacles*, Stanley G. Weinbaum describes "a movie that gives one sight and sound. Suppose now I add taste, smell, even touch ... Suppose I make it so that you are in the story, you speak to the shadows, and the shadows reply, and instead of being on a screen, the story is all about you, and you are in it. Would that be to make real a dream?" Such a fantastical description may sound familiar. Virtual reality (VR) describes technology that allows a user to experience an entirely virtual world. Though VR is often seen as a cutting-edge technology of the modern day, its core concepts originated much earlier. In fact, *Pygmalion's Spectacles* was written in 1935, and some of the foundational VR technology was created even earlier. With VR existing for decades, what has prevented it from fully catching on? To answer this, we'll have to start at the very beginning.

G. Kao et al., *Turning Silicon into Gold*,
https://doi.org/10.1007/978-1-4842-5629-9_19

The Origins of Virtual Reality

Fundamental elements of virtual reality can be traced back to technology invented as early as the 1830s. In 1838, the English scientist Sir Charles Wheatstone explained the concept of stereopsis, the perception of depth and three-dimensional space that is produced by our brain through the combination of visual stimuli from the left and right eyes.

From this understanding, Wheatstone invented the stereoscope, a device that displays a two-dimensional image for each eye that, when viewed, creates the illusion of a three-dimensional scene. To create this effect, two images are taken of the same scene at different points. The brain then processes and combines these images to create a single image with a sense of depth, thus giving the 3D effect. Indeed, as soon as humankind began to understand stereopsis, they worked to exploit it and create visual illusions to mimic reality. This discovery of stereopsis is fundamental to even modern-day virtual reality, which depends on creating the illusion of depth using two-dimensional screens.

Jumping forward to 1929, Edward Link, an American entrepreneur, invented the "Link Trainer," the world's very first flight simulator.[1] The "Link Trainers," or "Blue Boxes," were entirely electromechanical and would simulate various flight conditions and orientations. The student would sit inside the device, which looked like a miniature airplane. During World War II, tens of thousands of men and women received initial training through these devices. They also paved the way for a vital application of virtual reality—training simulations and other industrial applications.

Moving into 1956, we saw the beginning of another hugely popular application of virtual reality—the entertainment industry. This now multibillion dollar industry had its humble beginnings in 1956 with the Sensorama.[2] Invented by Morton Heilig, this highly immersive, multisensory machine looks much like a modern-day arcade machine with a hood around the user's head. It included a stereoscopic color display (thanks to Wheatstone!), odor emitters, stereo sound, and a vibrating seat and could even mimic atmospheric effects like wind. Even though viewers could not interact with the stories on the screen, the Sensorama helped introduce a future of immersive entertainment.

As you may have noticed, many of the early virtual reality devices were large hunkering machines that a user sat inside of. In 1960, this all changed when Heilig patented the Telesphere Mask,[3] the world's first head-mounted display (HMD),

---

[1] www.asme.org/wwwasmeorg/media/resourcefiles/aboutasme/who%20we%20are/engineering%20history/landmarks/210-link-c-3-flight-trainer.pdf
[2] www.mortonheilig.com/SensoramaPatent.pdf
[3] www.mortonheilig.com/InventorVR.html

bringing us much closer to the modern day conception of virtual reality devices. In fact, the patent drawings of the Telesphere Mask look uncannily similar to the virtual reality headsets of today. Later that decade, Ivan Sutherland and Bob Sproull brought virtual reality even closer to modern day by creating the Sword of Damocles,[4] the first HMD system to interface with a computer.

Looking through this history, the basic concepts of virtual reality began much earlier than one might have originally thought. The beloved idea of the hand-held VR headset started to take shape in the 1960s. If the basic concepts of virtual reality have existed for so long, why is it seen as such a hot trend now? With such a long history, why have previous attempts to popularize VR failed?

Virtual Reality in the 1980s and 1990s

Virtual reality had its first introduction to the mainstream during the 1980s, when leaps in technology brought virtual reality into the public eye. Research by NASA and universities such as MIT provided major technological advancements, particularly in hardware. After seeing major advancements in the 1970s, the field of computer graphics was beginning to commercialize. This is important because immersive VR requires both robust hardware and high-quality graphics.

After this technological boom, virtual reality saw widespread commercial release during the 1990s. Companies, particularly those in the entertainment industry, were rushing to capitalize on the mounting hype surrounding VR. This became the first golden age of virtual reality. Many large names in entertainment, such as Sega, Nintendo, and Disney, began to invest in driving mass adoption of virtual reality for consumer entertainment. In early 1991, Sega was hurrying to develop an at-home virtual reality headset and announced their plan to release the Sega VR headset. A couple years later, Nintendo too wanted a piece of the action and released the Virtual Boy in 1995. In 1998, Disney announced DisneyQuest, an indoor interactive theme park, and pumped $90 million[5] into the development of the attraction which included rides using head-mounted displays such as Aladdin's Magic Carpet Ride and Ride the Comix.[6]

---

[4]http://cacs.usc.edu/education/cs653/Sutherland-HeadmountedDisplay-AFIPS68.pdf
[5]www.thedp.com/article/1998/12/disney-plans-phila-theme-park-complex
[6]www.polygon.com/features/2018/10/18/17888722/disneyquest-disney-vr-closed

Yet with all this excitement about commercialized virtual reality, these companies saw only failure after failure. The launch of Sega's VR console was completely cancelled due to issues with motion sickness and headaches during user trials. Though Nintendo was able to release the Virtual Boy, only a measly 770,000 units were sold, winning it the title of Nintendo's worst selling console of all time (for reference, the Game Boy had sold 40 million units).[7] Failure of the console was attributed to high prices and unimpressive technology such as lack of portability, a monochrome display, and an unimpressive 3D effect. Even Disney's efforts were unsuccessful with the Chicago branch of DisneyQuest closing two years after its founding, and zero locations still open to date.

Unfortunately, the technological reality of VR in this era could not meet the lofty expectations of consumers or businesses. VR headsets were much too bulky. Computers were too slow, and graphics were not high enough resolution. People consistently got headaches and motion sickness, eventually even coining the term "virtual reality sickness." Thus, after this period of hype, both public interest and investor interest died down during the 2000s, marking the end of the first true rise of virtual reality.

Current-Day Virtual Reality

After a period of quiet, many companies both large and small started to revitalize the virtual reality industry, with interest in virtual reality once again rising to a relative high (particularly investment interest). In fact, the valuation of global augmented and virtual reality start-ups totals over $45 billion as of 2019.[8] Tech giants such as Amazon, Apple, Facebook, Google, Microsoft, Sony, and Samsung all have dedicated augmented and virtual reality research groups. Though there are many conflicting arguments about the future success or failure of VR, there is no denying the current boom in interest. In the following sections, we will explore the applications of modern-day virtual reality.

Note While modern-day conversations about VR often include discussions of augmented reality (AR), which superimposes virtually generated images on the user's view of the real world, the following sections are intentionally focused on virtual reality and its completely immersive experience.

---

[7] www.fastcompany.com/3050016/unraveling-the-enigma-of-nintendos-virtual-boy-20-years-later
[8] https://techcrunch.com/2019/10/18/vr-ar-startup-valuations-reach-45-billion-on-paper/

VR in Gaming and Entertainment

By far the primary driver of the VR industry nowadays is gaming and entertainment. Many of the biggest breakthroughs in virtual reality technology are out of gaming companies. In 2010, Palmer Luckey designed the first prototype of the Oculus Rift, a virtual reality headset designed to be inexpensive for gamers. In 2013, Valve Corporation, a multibillion dollar video gaming company, discovered technological breakthroughs in displays that allowed VR content to be displayed lag- and smear-free. In 2015, HTC, a Taiwanese electronics company, partnered with Valve to create the HTC Vive.

As a concept, it should not be surprising that VR is extremely attractive in the entertainment space. Even decades before now, the idea of being able to lose yourself in a dreamworld had tremendous appeal. VR provides a highly immersive experience with a level of interaction above current offerings. New forms of content also arise because of this new medium where storytelling can become highly interactive and engaging. So why hasn't this dream become a reality?

Currently, one of the largest barriers is accessibility, with the average price of VR headsets sitting in the couple hundreds of dollars. This price is hard for consumers to pay when virtual reality has not yet truly delivered on its value proposition—creating a uniquely immersive experience that only VR can provide. There are also still some inconsistencies in the quality of experiences that consumers expect, particularly in challenges with gesture and movement controls. In order to gain mass adoption in gaming and entertainment, virtual reality companies need to discover a reliable and consistent user experience that is truly immersive.

Mobile VR

With the recent rise of mobile technology, companies such as Google and Samsung have also invested many resources into mobile virtual reality, specifically phone-based VR. Hoping to address accessibility concerns, companies had high hopes in the adoption of mobile VR. In 2015, Samsung partnered with Oculus to create the Gear VR, a headset compatible with the flagship Samsung Galaxy smartphones, successfully selling 7.8 million units.[9] In 2016, Google released the Google Daydream View, a lightweight fabric headset which holds your smartphone, and shipped over 2 million units.[10] Yet even with much more impressive numbers than were seen in the 1980s and 1990s, phone-based VR was not here to stay, with both Google and Samsung

---

[9] www.theverge.com/2019/2/22/18234594/samsung-gear-vr-unpacked-2019-mobile-headset-future
[10] www.statista.com/statistics/752110/global-vr-headset-sales-by-brand/

shutting down their respective product lines in 2019.[11] The technological disadvantages of mobile VR were becoming too obvious, with intensive 3D applications quickly draining precious phone battery. People also reported friction when being disconnected from their phones because of the necessary complete immersion of virtual reality, displaying a lack of product fit within the market. Thus, within a short period of time, mobile VR both rose and fell.

VR for Training

Virtual reality has gained traction as a training tool in a variety of industries. In fact, Walmart has used virtual reality headsets to train more than a million of its employees and have seen astonishing results, reporting a 10 to 15 percent increase in test results. In particular, Walmart has praised VR for helping instill confidence in its employees, saying that "because the effect of VR training is like an experience in real life, associates have the freedom to make mistakes and learn by 'doing,' all while in a safe environment."[12] This method is also cost-effective, engaging, and provides valuable insights and metrics on realistic training scenarios.

For these very same reasons and more, the US military also uses virtual reality for military training, whether it is for flight simulations or medical training. Overall, although virtual reality has been used effectively for training in a variety of industries, it has currently only seen usage with a small percentage of potential adopters. Even with positive response from current users, we have yet to witness the mainstream adoption of VR in this use case.

VR for Communication

Four years after Oculus was founded, Facebook acquired the company for a whopping $3 billion.[13] The thinking behind the acquisition was in line with Facebook's overall mission, with a focus on using VR to facilitate human connection and help people come together in Social VR. In 2019, Facebook announced the launch of a new massively multiplayer online VR world called

---

[11] www.theverge.com/2019/10/16/20915791/google-daydream-samsung-oculus-gear-vr-mobile-vr-platforms-dead

[12] https://corporate.walmart.com/newsroom/innovation/20180920/how-vr-is-transforming-the-way-we-train-associates?irgwc=1&sourceid=imp_wiPwkD-SpyxyJT0TwUxOMo34iUknzaDzD3xP-Rg0&veh=aff&wmlspartner=imp_357605&clickid=wiPwkDSpyxyJT0TwUxOMo34iUknzaDzD3xP-Rg0

[13] www.businessinsider.com/facebook-zenimax-oculus-vr-lawsuit-explained-2017-2#oculus-co-founder-palmer-luckey-also-spoke-under-oath-he-disappeared-from-the-public-eye-following-revelations-about-his-political-ties-in-september-2016-but-has-since-reappeared-with-a-new-company-9

Horizon. As Facebook CEO Mark Zuckerberg explains, "What AR and VR do is deliver a sense of 'presence,' where you actually feel like you're there with a person, it's a really deep connection."[14] Though Facebook is investing heavily in this product area, the success of social VR is still very much an open question.

Overall, there are still a lot of exciting developments happening in the virtual reality space. In addition to the areas mentioned above, the applications of VR are also being explored in marketing, education, medicine, and more. Yet even across industries, for now, much of virtual reality usage is stuck with early adopters, and failing to go widely mainstream. While previous failures of adoption were attributed to technological failings, nowadays VR technology is developing rapidly, including developments in eye tracking and haptic feedback, and this is no longer as much of a barrier. With applications in consumer and enterprise, now might be the time to see if virtual reality companies can find the right product market fit, successfully acquire users, and create truly sticky and delightful products.

Takeaways: After analyzing the history of virtual reality through its rises and falls, it is hard to confidently say that the current interest in VR is here to stay. What we can say for sure is that previous technological issues are no longer a barrier to adoption, and the greatest obstacle in the current era of VR is finding the proper product market fit and growing past early adoption.

Questions

1. What other technologies have experienced similar rises and falls in adoption? How did they manage to gain critical mass?

2. In which industries do you think VR can become the most successful?

---

[14] www.cnet.com/features/facebooks-zuckerberg-isnt-giving-up-on-oculus-or-virtual-reality/

Society

Chinese Women in Tech

Finding More Success Than Their American Counterparts

"Women hold up half the sky," Mao Zedong famously declared in 1968, a diamond in the rough and violent legacy he left behind. Two decades prior to Mao's declaration, the Communist Party had assumed power in China and began the process of reshaping the country's ideological consensus on economic production and cultural development. Complete with a new name (the People's Republic of China), this reformed nation was to be a more balanced, more equal society lead by a dictatorship that would impress this creed on the Chinese people, no matter the cost.

In the process, the Communist Party under Mao committed human atrocities left and right, most notably the torture and public humiliation of those who opposed the "Great Leap Forward," Mao's five-year plan for the nation. However, in pushing for a more equal society, Mao's agenda also championed the abolition of traditional gender roles. This was particularly true when it meant fostering a more productive society motivated by propaganda in the absence of economic incentive. He wanted women to contribute to the labor

© Griffin Kao, Jessica Hong, Michael Perusse and Weizhen Sheng 2020
G. Kao et al., *Turning Silicon into Gold*,
https://doi.org/10.1007/978-1-4842-5629-9_20

force, and while this was still problematic in that women were expected to work and raise a family at the same time, this Maoist tenet was one of the few that left a positive impact—including a more equitable Chinese tech industry decades later.

Chinese Tech

Mao's dogma paved the way for gender equity strides in the workplace that have outpaced those in China's domestic sphere, and several decades later, Chinese women in the industry actually tend to fare better on average than their counterparts in other Asian countries like Japan, South Korea, and India.[1] In fact, in the tech industry, Chinese women seem to find even more success than their *American* peers, running counter to the idea of superior Western progressivism. According to *The Atlantic,* around 54% of US tech companies have women in C-level jobs (CEO, CFO, CTO, etc.), while almost 80% of Chinese tech companies do.[2]

Mao's cultural revolution seems to have planted the seeds that coincided nicely with the advent of the tech sector to create a perfect storm in which Chinese women in tech seem to be empowered at all levels, not just at the C-suite. 34% of US tech companies have at least one female director on their board, while 61% of Chinese tech companies do. And among the top US venture firms, only 10% of investing partners are women compared to 17% of investing partners at Chinese firms, according to Bloomberg.[3] Moreover, the Chinese government estimates that over half of new Internet companies are founded by women (although this estimate should be taken with a grain of salt given the source).

Prior to Mao's ascendency, China maintained a long history of female oppression, originating in part from Confucius' beliefs that women should be wives first, subservient to their husbands. The misogynistic doctrine was reinforced for centuries by education that focused on teaching women obedience instead of intellectual development, as well as a strong tradition of filial piety and patriarchism. Thus, by the onset of the tech industry at the end of the 20th century, many traditional sectors such as real estate and manufacturing were infused with sexist ideals, and Chinese women began to perceive tech, in comparison, as having much lower barriers to entry and more women-friendly workplaces.

---

[1] https://chinapower.csis.org/china-gender-inequality/
[2] www.theatlantic.com/technology/archive/2017/11/women-china-tech/545588/
[3] www.bloomberg.com/news/features/2016-09-19/how-women-won-a-leading-role-in-china-s-venture-capital-industry

A Deeper Dive

At a more granular level, there are several determinants instrumental to the success of Chinese women in tech, most of which stem from a larger cultural shift. In particular, the contribution of education in China cannot be understated. After Deng Xiaoping succeeded Mao, he began to implement policies for economic modernization focused on "reform and opening."[4] As part of this, Xiaoping worked to improve China's relationship with foreign powers and industries, sending students overseas to study in countries such as the United States on the government's dime. These policies prompted an era of unprecedented economic and social growth in China and gave women, now granted access to education, exposure to the innovative spaces in other countries. Indeed, many of China's defining female tech leaders like Jean Liu and Jane Sun were educated in the United States. Liu, the President of ride-sharing company Didi Chuxing (valued at $57.6 in 2019),[5] attended Harvard, while Sun, CEO of Ctrip, the largest online travel agency in China, attended the University of Florida.

Xiaoping also introduced the one-child policy in 1979, limiting Chinese parents to having a single child, which de-emphasized the responsibility of putting motherhood first that was heavily impressed upon Chinese women. To be clear, this policy that resulted in countless missing Chinese girls and forced sterilizations and ninth-month abortions was in so many ways detrimental to women. But for the specific demographic of girls born in major Chinese cities after 1980, their educational and career prospects began to improve.[6] To complement the one-child policy, traditional Chinese values surrounding family encouraged tight-knit households where grandparents were oftentimes quite involved in the upbringing of the grandchildren. Thus, the shared load of parenting allowed women more freedom in their professional lives and meant family structure became a tool they could leverage in the pursuit of their career aspirations.

Although less measurable, there appears to be a strong community of women in tech that provides a network to support and affirm female participation in the industry. Sun of Ctrip has spotlighted that many of the women at Chinese Nasdaq-listed companies, including the CFO of Alibaba, are "good friends" of hers.[7] More broadly, 63% of Chinese tech companies say they have programs in place to increase the number of women in leadership positions, compared

---

[4] www.washingtonpost.com/news/monkey-cage/wp/2018/12/19/40-years-ago-deng-xiaoping-changed-china-and-the-world/
[5] www.cnbc.com/2019/05/14/didi-chuxing-2019-disruptor-50.html
[6] Mei Fong, *One Child: The Story of China's Most Radical Experiment* (Houghton Mifflin Harcourt, 2016).
[7] https://money.cnn.com/2014/08/18/technology/china-women-tech/

to the less than 34% of US tech companies, which indicates that female leaders are committed to cultivating an environment that makes it easier for other Chinese women to succeed in tech.[8]

A Multilevel Solution

It should be noted that the Chinese tech industry still has a lot of room for growth in gender equity with its continuing issues of pay disparities and workplace harassment. However, China seems to be in a more progressive place than other countries on this issue, in a world where 21 of the 56 self-made female billionaires in 2017 are Chinese, according to *Forbes* (Zhou Qunfei, the richest, made her fortune on smartphone screens).[9] Many of the factors that contributed to these strides in gender equality are not replicable and/or shouldn't even be attempted, from a violent cultural revolution to invasive child-rearing policies. Still, for those in the tech industries (and governments) of other countries, what we can learn from China is that the solution to gender discrimination will likely come from multiple levels. Starting from the top, we need more equitable policies—in particular those that provide access to education and a level playing field in the workplace—and female representation among leadership. At the bottom, there needs to be foundational systems in place for women to feel heard and enabled to contribute in their day-to-day work. Only then can we empower women in tech to truly hold up half the sky.

Takeaways: Women in Chinese tech seem to be finding more success than their American counterparts. Reasons for this may include Mao's cultural revolution coinciding with the development of the tech sector, educational policies, advantageous family structure, and strong female peer networks. These factors suggest that progress in gender equity within the American tech industry will be driven by institutional changes made at the C-suite level. However, we can also note that such policies often go hand in hand with action at the individual level.

---

[8] www.theatlantic.com/technology/archive/2017/11/women-china-tech/545588/
[9] www.forbes.com/sites/chloesorvino/2017/03/08/the-worlds-56-self-made-women-billionaires-the-definitive-ranking/#4f9e898168a2

Questions

1. For prospective female tech employees, what are some barriers to entry in America? These could include hiring practices, workplace culture, leadership representation, etc.

2. What are some policies that employers can adopt that approximate or incorporate the ideas behind the Chinese family structure that facilitates women succeeding in the workplace?

3. What are actions that individual male employees can take to be a better ally to women? Some examples could include participating in diversity trainings, ensuring all meeting participants get a chance to contribute, etc.

Mental Health Algorithms

How Posting Facebook Statuses Today Might One Day Save Your Life

The moment you uploaded your first Facebook profile picture, you took a critical leap in establishing an online version of your identity, telling the world, "*this* is what I look like" (at least, from your most flattering angle). As the years went by, your pictures would rack up more likes—and eventually, heart reacts—as that digital representation of you grew to take up more and more storage on Facebook's massive servers. Your profile quickly became an online repository of everything you've commented and posted, all the things you've liked, and all of the messages you've sent and received, providing a fuller picture of who you are as a person and how you'd like your immediate world (and advertisers) to see you. But beyond the binary and targeted advertising, your Facebook profile has morphed into something even more special—a digital representation of your *psyche*. As we've we all afforded social platforms like Facebook more and more significance to our actual person and the way we live our lives, we've grown their ability to provide mental snapshots of ourselves through time.

© Griffin Kao, Jessica Hong, Michael Perusse and Weizhen Sheng 2020
G. Kao et al., *Turning Silicon into Gold*,
https://doi.org/10.1007/978-1-4842-5629-9_21

Opportunity for Intervention

While this social shift toward online personas was occurring, suicide grew to be the 10th leading cause of death in the United States as of 2016, and it was estimated by the World Health Organization (WHO) to meaningfully impact the lives of over 130 people with each incidence. WHO also reports that 90% of suicides can be attributed to *chronic* mental illness, which has made suicide screening, prevention, and ideation research a significant responsibility and immediate problem area for the psychiatric community. Finding new venues for assessing and preventing high-risk individuals from committing suicide is therefore of pressing interest to the medical community, and social media has proven to be an abundant source of *personalized* data.

While other, more traditionally physical forms of illness—such as heart disease—are more easily diagnosable across all individuals, the medical community has struggled to find an accurate way to screen the public for suicide risk. Could the advent of personalized data on web sites like Facebook, which has captured about 70% of the US population as daily users, be a solution? The answer, so far, seems to be "yes."

While social media has grown to become an important part of our society and our concepts of self, the existence of online, data-rich personas for billions of people offers an exciting space for data scientists and psychologists to collaborate. Both parties have the shared goal of predicting which individuals are at a high risk of committing suicide so that intervention and treatment can be offered. But what evidence and methods exist that make social media data a valid predictor for mental health?

Creating a Mental Health Algorithm

To start, the advent of natural language processing (NLP) techniques has been an exciting area of growth in the field of machine learning (ML). One such technique is sentiment analysis, which can take an input string of text and assign it a variety of semantic and emotional valences, hopefully corresponding to how a human might interpret that text in its natural setting. For instance, text can be assessed to check its sentiment for how likely it is to be indicative of suicidal thoughts and ideation. Expanding upon this, by looking at individuals' streams of text over time through messages, tweets, status updates, and the like, one can hopefully paint a full and accurate picture of the average progression of that person's mental health. ML models can then be created to predict the candidacy of suicidal behavior and ideation, which can then in turn be used to flag the given individual for intervention and even personalized treatment based on the model's findings.

There are a variety of methods to create such models, and all have their own technical, ethical, or psychiatric faults. One such model was created from a study that pulled social media data from people who had made at least one unsuccessful suicide attempt. Then, a timeline was made to sync a given person's social media data to their mental health at certain points, the ideation period leading up to the suicide attempt, the day of the suicide attempt, and the period following the attempt and interventional treatment. By aggregating this data, the ML approach of unsupervised learning (an approach that finds previously unknown patterns/labels in data) was used to create a model that could then predict risk in others.

Another method for processing social media data into screening models and evaluators involved using Twitter to build a vocabulary of suicidal terms and phrases. Given the nature of Twitter as a character-limited, text-based platform, researchers were able to systematically parse tweets that showed signs of suicidal sentiment before reviewing them manually for human filtering. By combining human knowledge with an ML model's capability to learn, a complex neural net was created that could predict tweets that were indicative of high risk of suicide with incredible accuracy. One objection made to this technique was that "exhaustive sentiment analysis" would never be a foolproof method for screening, given that false positives (falsely labeling someone non-suicidal as suicidal) and false negatives (falsely labeling someone suicidal as non-suicidal) each have their own have negative consequences, yet fine-tuning the model seems to only shift the amount of negatives there are between the two. This lack of annotated data in raw information such as a spurious tweet has also posed a problem for other researchers using similar methods. It seems like the abundance of data for social media is both a blessing and a curse; filtering out important information that would be indicative of poor mental health is a challenging endeavor. Fortunately, however, these models do seem to function as intended when the person is at a high point of likelihood to commit suicide, a time where immediate intervention is also most wanted.

Another study that tested social media's validation as a data set presented psychiatrists with all of an individual's Facebook statuses to assess suicide risk. It found that Facebook statuses were sufficient enough to accurately assess someone's depressive symptoms, given that psychiatrists generally agreed with in-person assessments of an individual as they did their Facebook history of statuses. An important understanding and revelation that this study provides comes from the relevance of Facebook in 2011 vs. Facebook today. The population of people that were being tested in this 2011 study were current college students, and Facebook has since declined as a status-posting web site in the college-aged demographic. We must then ask how much authority we would be willing to give social media data screening if its popularity and usage is liable to change. If we were to build out and incorporate screening by social media technologies and then usage was to change, we might lag behind optimal screening methods.

As data becomes a tool and resource in more areas of study within psychology, a larger range of qualitatively classified phenomena such as mental illness will be able to be studied with quantitative correlations in information such as social media data. To build out the available methods of analyzing text for signs of mental illness with an emphasis on depressive symptoms, psychiatrists collaborated with data scientists to create a strong model of words and phrases that can predict whether or not a series of Twitter posts have high-risk suicidal intent. Infusing the classification requirements that psychiatrists use to assess patients with into the classification algorithms within the model strengthens the validity of these automatic risk detection systems.

Additionally, these suicide algorithms have sought validity by comparing their results directly to clinical data in validation tests. For instance, Chinese microblog users were given an examination to place them on the Chinese Suicide Probability Scale (SPS), and their microblog data was assessed by existing models. Results showed that these models were also able to match those with high SPS values to highly likely candidates for depressive and suicidal conditions.

Big Tech, Big Data, Big Responsibility?

Large tech companies such as Facebook/Instagram, Twitter, and Google are situated at an interesting place within the discussion of using data for mental health screening purposes. They are the owner of the data they collect given their position as hosts of the social media platform (or information platform in Google's instance). As applied to mental health, it is clear that the data these companies have is incredibly rich for assessing the mental health of its users. What to do with the data and how to act on any findings quickly becomes a challenging topic with questionable motives. Should these tech giants see themselves as socially obligated to screen and initiate prevention for their users at risk of suicide if the data is there ripe for use? Should these companies be banned from overstepping into the mental health assessments of their users given a lack of domain knowledge and/or privacy concerns?

Each company is still developing its own answers to these questions while trying to avoid being labeled as invaders of *public privacy*—an oxymoron stemming from the fact that there are private and personal conclusions to be drawn from publicly viewable data.

In 2018, Facebook created what it calls its "proactive [suicide] detection algorithm." The most common use case for the algorithm is as follows. A given Facebook user begins a live video, and the comments from that person's Facebook friends are evaluated. If the comments are deemed to be high risk of a suicidal video (comments that plead the person to stop, comments that show love and sadness sentiments), then the video is ended, flagged, and is shown to a team of people working for Facebook. If there is sufficient evidence of danger to the person, then the authorities are called. Facebook called the police on 3,500 people in the United States using this technique and would not release information as to how successful those prevention attempts were. In a press statement on the new technology, Facebook formally announced the new feature to its platform: "based on feedback from experts, we are testing a streamlined reporting process using pattern recognition in posts previously reported for suicide. This artificial intelligence approach will make the option to report a post about 'suicide or self-injury' more prominent for potentially concerning posts like these." This model employs the natural language processing techniques that were discussed earlier, as well as the human–computer shared interaction of using the algorithm as a first level of screening before escalating to a human source.

In this instance, Facebook might feel obligated to act as mediator and escalator in high-risk scenarios due to pressure from the public and fear of bad PR. Pulling suicidal videos with the advent of live streaming is something Facebook has only recently begun to do since 2018. What they should do (or what they are doing, privately) with the rest of their users' data to assess mental health is still a matter of contention.

Google is another company with an abundance of user profile data. Should it be legally or socially obligated to do anything when users search for things related to suicide techniques? The answer is also unclear, and for the meantime, it appears to be "no" given public opinion surrounding the invasion of privacy. Google has settled for simply listing suicide hotline links and information as the top search results whenever a suggestively suicidal query is made.

These large tech companies currently choose to selectively prioritize their privacy with regard to their users' data when it comes to mental health screening, understandably, given that it is an invasive, personal, and somewhat taboo label that feels reminiscent of a dystopian novel. As technology becomes more pervasive in life and as psychiatric comfort with using big data to screen people grows, however, it is worth considering the possibility of these companies becoming obligated to provide data to protect the health of their users.

Generalizing Suicide Risk Predictions to Other Mental Diseases

There is nothing particularly unique to screening for suicidal risk using social media data. Rather, the entire technique is a new frontier in and of itself in the realm of assessing and diagnosing social media users with risk levels and mental illnesses. Thus, these same tactics could be applied to other areas of concern, such as eating disorders, and they have recently been applied to make models that help screen for PTSD.

Additionally, techniques of classifying mental illnesses using machine learning position the entire task of classifying diseases in a new light. Part of the beauty of ML models such as neural nets is that they find connections and clusters of data that sometimes aren't even apparent to human intuition. There have already been examples of ML models creating their own subgroups for people with the overarching label of major depressive disorder (MDD), suggesting an increased need for nuance toward the existing classification schemas. This could be potent for creating new clinical cutoffs for various levels of depression to better suit individuals toward personalized medicine. Allowing for an individual with MDD to then have their social media data analyzed could make this personalization a reality.

Amidst all the uncertainties of this new technology, several open questions remain: Should companies like Facebook be obligated to analyze this data for the sake of public health, or should it be prohibited as an invasion of privacy? These businesses have found themselves navigating the problems of social well-being before—is this an opportunity to make amends? Data is by no means the panacea to solving mental health issues, but social media has presented itself as a way of correctly assessing risk at scale.

Takeaways: Companies like Facebook with enormous amounts of user data have been proven to be able to predict user suicide risk with some reliability. There are many privacy concerns surrounding this possibility, and also undefined responsibilities for tech companies, which have no precedent for serving as public health moderators.

Question

1. Do companies like Facebook have a responsibility to use their user data for public health and well-being? Consider ethical, governmental, legal, and corporate viewpoints.

I Made It; It's Mine

Why Tech Founders Have an Irregular Amount of Control Over Their Public Companies

Zuckerberg, Gates, Jobs—all founders that have been given celebrity status as the face of companies that have changed the world with technology. Gaining traction as household names and growing their visions and companies took considerable outside financial and operational investment; despite this, these founders and others like them have managed to maintain a massive amount of control over their companies through a variety of methods.

Shares and Voting Rights

In order to understand how the elusive idea of "company power" can be quantified, let's review what shares of a public company offer their shareholders.

When a company IPOs (meaning, performs an initial public offering and begins to sell its shares on the stock market), they are allowing public investors (i.e., you and me) to buy a piece of their company in the form of shares (see Chapter 13 for more on IPOs). Holding shares in a company traditionally

G. Kao et al., *Turning Silicon into Gold*,
https://doi.org/10.1007/978-1-4842-5629-9_22

means two things—(1) the shareholder gets profits and (2) the shareholder gets voting rights within the company that are proportional to their ownership. As the profit aspect of shareholding has been what stocks are typically considered for, the voting rights component has gradually faded away as tech founders negotiated for more and more control as their companies went public.

Voting rights offer their shareholders a say on decisions for the company, such as the product direction, electing a new member to the board of directors, or whether or not to fire a CEO. In the general case, one share of a company equals one vote. Anyone who owns shares can vote at shareholder meetings with the company's board of directors or by proxy through the mail, Internet, or phone.[1]

Recently however, the "one share, one vote" structure has been diluted by companies offering multiple classes of shares. Some classes offer voting rights, some offer none, and some offer "super-voting" privileges, where one share is equivalent to multiple votes. Let's take a look at Google's parent company, Alphabet, for an example:

- **Alphabet Class-A shares:** one vote per share, publicly traded as "GOOGL"

- **Alphabet Class-B shares:** super-voting shares, 10 votes per share, held by founders and early investors, not publicly traded

- **Alphabet Class-C shares:** no voting rights, publicly traded as "GOOG"

Alphabet's Class-A shares are referred to as "common stock," where shareholders own and trade some voting rights, the value of the company, and its profits. Their Class-C shares are purely traded for profit sharing. Class A and C shares typically hover around the same value and were created after a **stock split**, where a company duplicates the number of shares it has to make owning and trading the stock seem more affordable. The negligible price difference despite the disparity in voting rights seems to prove that the average investor isn't too worried about voting rights.

To founders, however, maintaining control over voting rights is an imperative, which is why we see Alphabet's peculiar Class-B shares. This class of shares was constructed when Google first went public, and their "super-voting" power allows for founders Larry Page and Sergey Brin to retain a majority of

---

[1] www.asyousow.org/vote-your-shares

Alphabet's voting power despite owning less than a majority of the total company's stock. By constructing the classes of shares like this, the founder duo was able to allow for as much public investment as possible without watering down their control over the company. In a 2012 founders' letter, Page and Brin preemptively addressed concerns that would arise surrounding this voting system:

"We recognize that some people, particularly those who opposed this structure at the start, won't support this change—and we understand that other companies have been very successful with more traditional governance models. But after careful consideration with our board of directors, we have decided that maintaining this founder-led approach is in the best interests of Google, our shareholders and our users. Having the flexibility to use stock without diluting our structure will help ensure we are set up for success for decades to come."[2]

The notion of a founder-led approach is clearly core to Alphabet, as well as many other tech companies that both preceded and would follow in its wake.

Next, we'll explore if this trend holds up across other public tech companies and examine voting share power as a primary way for tech founders to exert control, even amidst personal controversy with their board or the public.

Founders, Their Companies, and Voting Shares

The following are several influential, public tech companies, their founders, and a snapshot summary of how the founders have control over their companies in the form of voting rights or other measures (all data are from 2019):

Alphabet

- Founders Larry Page and Sergey Brin manage 51% of Alphabet's voting power, yet own 13% of its stock.

- Page and Brin are known to blanket-vote "no" while absent from some board meetings and have voted against protests against the model, which were intentionally moot attempts to move the company to a one-vote-one-share model.[3]

---

[2] www.sec.gov/Archives/edgar/data/1288776/000119312512160666/
d333341dex993.htm
[3] www.vox.com/2017/6/13/15788892/alphabet-shareholder-
proposals-fair-shares-counted-equally-no-supervote

Facebook

- Founder Mark Zuckerberg owns approximately 20% of Facebook's total shares across all classes, yet controls about 60% of the voting rights.

- Similar to Alphabet, Zuckerberg owns a majority of Facebook's Class-B shares, which have super-voting status of 10 votes per share.

Snap

- Founders Evan Spiegel and Bobby Murphy own a combined 96% of voting rights.

- In an unprecedented move, none of Snap's publicly traded Class-A shares offer any voting rights.[4] Its Class-B stock have one vote per share and are reserved for executives and early investors. Its Class-C shares have super-voting status with 10 votes per share.

SpaceX

- Founder Elon Musk owns 54% of SpaceX shares and 78% of its voting shares.

Spotify

- Founders Daniel Ed and Martin Lorentzen created and own a class of super shares that have only voting rights and no monetary value, giving them control over company voting.[5]

Tesla

- Founder Elon Musk owns only 22% of Tesla's shares (TSLA), but Tesla's bylaws require a supermajority (two-thirds) vote for any major company changes; thus, at least 85% of non-Musk shareholders would have to vote against something he disagreed with for it to pass.

It's a Tech Thing

But how specific is this founder-centric voting control to tech companies, rather than the founders of non-tech companies? As it turns out, it's very much a tech thing.

---

[4] www.vox.com/2017/2/21/14670314/snap-ipo-stock-voting-structure
[5] www.businessinsider.com/spotify-direct-public-offering-has-firewall-to-keep-founders-in-control-2018-2

As we've seen, the existence of multiple classes of stock in a company is a sign that founders are exhibiting voting control through special super-voting classes reserved for themselves and early investors. In 2019, half of tech IPOs had multi-class stock options, compared to only 15% of non-tech IPOs.[6]

It's evident that tech companies in the past few years have had more of these multi-class setups than non-tech companies, and perhaps this is what has contributed to the founder-centric image of big tech. Relatedly, the percentage of VC-backed public tech companies with supervoting shares has been strongly skewed toward tech companies. See Figure 22-1.

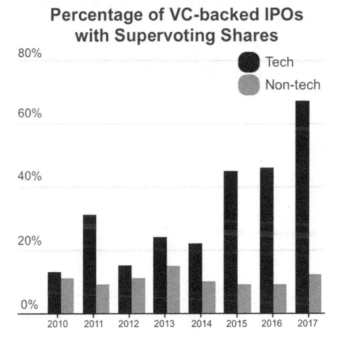

Figure 22-1. Voting Shares in Tech vs Non-tech IPOs[7]

[6] www.vox.com/2019/4/11/18302102/ipo-voting-multi-dual-stock-lyft-pinterest
[7] https://site.warrington.ufl.edu/ritter/files/2019/04/IPOs2018VC-backed-1.pdf

Ellen Pao, former CEO of Reddit and former venture capitalist at Kleiner Perkins, believed that VCs have been a major contributor toward the "founder-friendly Silicon Valley."[8] It seems that technology company founders are particularly determined to see themselves as the ones steering their company's vision and VCs have been willing to offer such control in the negotiating room.

Overall, tech company founders in particular have found a unique way of cementing themselves into the legal, economic, and bureaucratic frameworks of their companies through voting rights. Being able to lead their company through periods of intense scaling has made celebrities out of the founders who were able to navigate pitfalls successfully. Whether or not the landscape of venture capitalism remains founder-friendly as more and more tech companies flop during their IPOs (see Chapter 13) remains to be seen.

Takeaways: Tech company founders retain strong ownership over their public companies' decisions through special shares of stock with greater voting rights. Tech companies more than any other industry in recent years have seen this founder-centric control and leadership.

Questions

1. From a venture capitalist perspective, why might tech founders more than any other industry be retaining control over their company as more and more people invest?

2. What are some potential problems with having so much large company power concentrated within one or a few founders?

---

[8] www.wired.com/story/ellen-pao-founders-absolute-power-destroying-company-culture/

What to Expect When You're Expecting a Billion New Users

How Tech Companies Are Preparing for Users from Emerging Markets

Can you think back to a time when you didn't have the Internet? Not because you were on a camping trip or there was a power outage but because it just didn't *exist* where you lived? It's probably been decades since you've had those

© Griffin Kao, Jessica Hong, Michael Perusse and Weizhen Sheng 2020

G. Kao et al., *Turning Silicon into Gold*,

https://doi.org/10.1007/978-1-4842-5629-9_23

memories (if you're old enough to even have them), but in the coming decades, one billion new people from around the world—give or take—will be leaving those memories behind and joining the Internet.

Who Are the Next Billion Users (NBU)

These people have been dubbed the **next billion users** (NBU), and tech companies of all sizes are eagerly planning and strategizing for their "arrival." NBU countries are emerging markets with rapidly increasing trends in urbanization that are "primed for growth," in that much of the country is not online yet infrastructure exists to scale out affordable coverage in the coming

Table 23-1. Next Billion Users by Country

Country	Country population	Percent of population online
India	1,300,000,000	35.2%
Bangladesh	158,000,000	18.4%
Indonesia	261,000,000	54.8%
Nigeria	191,000,000	25.7%
Pakistan	205,000,000	15.6%
Philippines	104,000,000	55.8%

decades (Table 23-1)[1]

Overview of Urban Population Demographics

An estimated 68% percent of the world's population will live in urban areas by 2050.[2] Therefore, we can look to trends in urban growth to better understand what changes in technology availability and use cases might be occurring simultaneously. To better understand the significance of NBU, let's first examine urban population demographics from 2020 and projections for 2050 in Table 23-2.[3]

---

[1] https://media.bain.com/next-billion-internet-users/index.html
[2] www.un.org/development/desa/en/news/population/2018-revision-of-world-urbanization-prospects.html
[3] http://media.wix.com/ugd/672989_62cfa13ec4ba47788f78ad660489a2fa.pdf

Table 23-2. World's Largest Cities, 2020 and 2050

City	2020 Population	Projected 2050 population
Mumbai, India*	22.5 Million	42.4 Million
Delhi, India*	29.4 Million	36.2 Million
Dhaka, Bangladesh*	20.3 Million	35.2 Million
Kinshasa, Democratic Republic of the Congo	11.6 Million	35 Million
Kolkata, India*	5.8 Million	33 Million
Lagos, Nigeria*	21 Million	32.6 Million
Tokyo, Japan	37.2 Million	32.6 Million
Karachi, Pakistan*	15.7 Million	31.7 Million
New York, United States	19.4 Million	24.8 Million
Mexico City, Mexico*	21.2 Million	24.3 Million

*NBU countries

Urban life has traditionally been more conducive to Internet apps services due to the high population density providing an agar for niche markets; for instance, Uber and Uber Eats only work in relatively urban areas; Google Maps only has rich, location-based experiences in urban areas; and dating apps can offer wider pools of people to their picky users in urban areas where there are more users. The fact that more than two-thirds of the world will be living in urban environments coupled with the reality that the majority of the most populous cities in the world are in NBU countries means that technology companies need to start thinking about how to engage with NBU urban populations, or risk missing out on a huge pool of users.

Why NBU?

Other than a massive new pool of users, what makes investing in NBU markets worthwhile? At a glance, the major risks of moving into NBU markets are associated with the lower relative GDP for NBU countries as compared to countries like the United States (for context, the United States' GDP in 2017, 19.39 trillion USD; India's GDP in 2017, 2.597 trillion USD).[4] Put bluntly, there is less money to be made per user in NBU countries. Despite this, tech companies are betting on acquiring the next billion users to have a worldwide anchor in as many markets as possible. The future is hard to predict, and NBU countries' growth are one of the most promising statistics to make business decisions off of.

---

[4]https://data.worldbank.org/indicator/NY.GDP.MKTP.CD

A primary reason for focusing on NBU countries is that tech companies want to make first moves to penetrate these markets and find where they can succeed most efficiently. Companies will spend money trying to scale out their infrastructure and advertise in these areas in hopes of securing a foothold and one day making profits as they do in most Western markets.

In addition, many companies see NBU countries as a promising next step because they have existing products—they just need to scale them. Facebook won't be building a Facebook 2.0 for India; rather, it will be extending its infrastructure and polishing its product to fit the needs and norms of its Indian users. This is where most companies are expending effort in NBU: devoting software teams to best understand how users in NBU countries live and how to bring the products they've already invested in and which have proven successful to another billion users who will hopefully also find them useful.

Mobile Phones, Big Tech, and NBU

When you imagine the Internet, you're probably thinking of viewing a web browser on a monitor or laptop screen. But despite these established imaginations of what the Internet is, most of the next billion users will only interface with the Internet on mobile devices. As an extreme example, mobile phones in Kenya account for 99% of all Internet traffic in the entire country.[5]

The mobile experience will define the future of the Internet in NBU countries. Google and Facebook are working with NBU country telecom providers to subsidize the reach of at least *some* Internet connection; in 2020, approximately 70% of Southern Asia will still depend on 2G connections. Google and Facebook have to account for this when building out apps for NBU countries; every piece of data serviced between the phone and the company's networks must be important to the user experience, which might feel stripped or minimal in comparison to Western versions of the same apps.[6]

Google is making its push into NBU markets with an Android-first approach. The company which most people think of *as* the Internet offers external programs for Android developers called *Building for Billions* that teaches software engineers about designing apps that can work in low-data scenarios.[7]

---

[5] www.ictworks.org/mobile-data-subscriptions-account-99-percent-all-internet-access-kenya/#.W_W1VbaZNo4
[6] www.cgdev.org/sites/default/files/governing-big-techs-pursuit-next-billion-users.pdf
[7] www.cgdev.org/sites/default/files/governing-big-techs-pursuit-next-billion-users.pdf

Furthermore, Google has rebuilt its own native version of Android apps for low-data NBU countries, such as YouTube Go, a stripped-down version of YouTube that allows users to gradually download videos (at one of 400 free train station WiFi spots that Google provides in India, for instance) so that they can be watched later.

Small Tech and NBU: Emerging Market Start-Up Model

It's clear that large tech companies with existing capital have the time and resources to prepare for and hook emerging markets on their advanced products, but small tech companies are also moving fast to share in the volatility of NBU markets.

Several companies from Y Combinator, a prestigious start-up seed-level accelerator, have had business plans as simple as bringing an existing technology to a NBU market. For example[8]

- "TradeID" pitches themselves as "Robinhood for Indonesia."[9]

- "Coco Mercado" pitches themselves as "Instacart for Venezuela."[10]

- And "Tizeti" pitches themselves as a new "Internet Service Provider for Africa."[11]

It's clear that NBU countries present opportunities for start-ups as well as established products and brands. Applying validation from an existing success in Western markets to these NBU countries appears to be an emerging strategy for companies. Having founders with an understanding of these markets and the users in them will be a necessity for success, as the blind application of a successful product in one culture and city might not work for every pairing.

Takeaways: NBU countries are a focus for large and small tech companies alike. The next billion users will use existing products differently than Western markets and under different conditions. Understanding what those differences are is essential for development and successful market penetration.

---

[8]www.ycombinator.com/companies/
[9]www.gotrade.id/
[10]www.cocomercado.com/
[11]www.tizeti.com/

Questions

1. How should NBU countries respond to subsidized agreements between large tech companies such as Google and Facebook?

2. What's an example of a technology product that might not successfully transfer to an NBU market with ease?

Wisdom of the Crowd

Harnessing the Power of the Public

How many times has a teacher told you that Wikipedia is not a reliable source of information?

Wikipedia itself has a page titled "Wikipedia is not a reliable source," where it explains that literally anyone can edit its pages anytime. Yet, people still trust Wikipedia more than news that is written by trained journalists.[1] Even more telling, phrases from Wikipedia articles make their way into scientific papers, indicating that scientists read Wikipedia even if they don't admit it.[2] Although there is no single expert writing the articles on Wikipedia, these articles still seem to hold a certain weight in terms of trustworthiness and reliability. For some reason, we all seem to implicitly trust the collective authors of Wikipedia, subconsciously or not.

Wikipedia has successfully managed to pool together the knowledge of millions of Internet users and then subsequently built a foundation of trust among its huge reader base, all despite the lack of specific authorial jurisdiction. This phenomenon where the collective group of individuals is "more

---

[1] www.vice.com/en_us/article/ae37ee/in-defense-of-wikipedia
[2] www.sciencenews.org/blog/scicurious/wikipedia-science-reference-citations

© Griffin Kao, Jessica Hong, Michael Perusse and Weizhen Sheng 2020
G. Kao et al., *Turning Silicon into Gold*,
https://doi.org/10.1007/978-1-4842-5629-9_24

knowledgeable" than an individual expert is known as the wisdom of the crowd. The idea is that when taking the average over a huge crowd of diverse people and backgrounds, we will find the collective result is on par, if not better, than that by a single expert. By bringing in diverse opinions and thoughts from a group of people who each have their own private information and specialized knowledge, the aggregation of the group can have extraordinary power. Wikipedia is a prime example, where the collective group of Internet users pool their shared knowledge. The very reason why Wikipedia is considered "unreliable"—the fact that anyone can edit it—is also exactly why it is so trustworthy. We as users are free to put any information on the web site and, in turn, regulate the information other users post. This self-governance taps into the power of the wisdom of the crowd, where we trust that the thousands of Wikipedia readers check and regulate the online articles they read, ultimately collectively creating one of the world's most well-known and well-trusted encyclopedias.

Today's Online Crowd

In today's Information Age, characterized by the value placed on data and information, the wisdom of the crowd has become an increasingly popular concept. The Internet is built upon the idea of millions of online users contributing to content on the World Wide Web, creating a vibrant online community filled with information. Today, numerous companies take advantage of this supply of information and have found a multitude of applications for it.

One way companies have been able to harness the wisdom of the crowd is through crowdsourcing. Crowdsourcing is the term used to describe the sourcing of work and/or ideas from Internet users, and is a portmanteau of the words "crowd" and "outsourcing." By the nature of the people found on the Internet, crowdsourcing draws from a diverse group of people and theoretically provides quality work/answers as a result.

Dozens of examples of crowdsourcing web sites can be found online, such as fanfiction sites (i.e., FanFiction.Net), artwork communities (i.e., DeviantArt), and photo-sharing sites (i.e., Flickr). Many companies, including Yelp and Google, use user-generated content (UGC) to aggregate user knowledge and opinions on companies and organizations. UGC draws from countless users of the Internet to create content for a specific topic, usually involving ratings, reviews, and media sharing (i.e., photos)—think of online restaurant reviews and shared photos of food and menus. It's becoming an increasingly popular tool for companies, as it is an easy way to generate content, while simultaneously securing trust from its users. Studies have shown that 78% of users trust peer reviews more than branded content, and in addition, UGC is often communicated in an authentic way that drives meaningful conversations with other consumers.[3] There are countless examples of crowdsourcing online, and they all illustrate the endless potential and creativity that online crowds can achieve.

Idea Competitions

Companies can also leverage the crowd to improve upon their own products. Crowdsourcing can be an effective way to host "idea competitions," where companies learn from those outside their organizations.

The Netflix Prize is a well-known example of an open competition. Netflix announced the competition in 2006, promising $1 million to any team that could raise the accuracy of the company's recommendation engine by 10%.[4] They released a data set of over 100 million ratings of 17,770 movies from 480,189 customers, which 30,000 people downloaded in an attempt to rise and emerge victorious.[5] The difficult task of the contest prompted users to begin sharing their results and ideas online, giving other competing teams the opportunity to improve and expand on them and develop more and more complex algorithms. The contest lasted for three years, until finally, on June 26, 2009, a team finally reached the 10% mark.[6] Although Netflix didn't even end up using the final winning algorithm, some contestants believed that the technical advancements from the first year of the competition were definitely used by the company.[7] These advancements were a result of an open contest that encouraged collaboration among thousands. The competition prompted numerous other companies to enter the realm of data sharing and to draw inspiration from the crowd; the analytics competition platform Kaggle was founded in 2010 to further crowdsource data science problems, and Yelp and Zillow have similarly released huge data sets for the crowd to work on.[8]

Lego offers a less technical example that similarly draws inspiration from the public. Lego Ideas is a web site by The Lego Group that allows users to submit ideas for Lego products, rewarding the designer with 1% of net sales if successfully chosen to be produced.[9] Submitted ideas can gain supporters, and once they gain 10,000 supporters, they are considered for being one of the four designs made into an official Lego product.[10] As of 2018, the Lego Ideas

---

[3] www.insided.com/blog/the-value-of-real-peoples-opinions-why-user-generated-content-is-so-popular-in-these-6-categories/
[4] www.thrillist.com/entertainment/nation/the-netflix-prize
[5] Ibid.
[6] Ibid.
[7] Ibid.
[8] Ibid.
[9] https://ideas.lego.com/guidelines
[10] www.cnbc.com/2018/04/27/lego-marketing-strategy-made-it-world-favorite-toy-brand.html

online community had almost 1 million adult users.[11] It's a prime example of a company using the creative power and ideas of its own consumers to better develop products its users will want—in 2017, Maia Weinstock's "Women of NASA" was chosen and became an instant top-selling product on Amazon within 24 hours.[12] Not only does the competition ensure product market fit, it also provides Lego with free user research from its consumers in exchange for communicating that their opinions are valued. Companies are finding ways to distill the best ideas from the crowd and ultimately using them to provide value back for the crowd.

The "Pioneer of Crowdsourcing"

We end this chapter by exploring one of the best-known names in crowdsourcing who is often dubbed a crowdsourcing pioneer: Luis von Ahn. von Ahn refers to crowdsourcing as "human computation"[13] and, as we will explore, has shifted the crowdsourcing paradigm more than once within the span of a decade with reCAPTCHA (released 2007) and Duolingo (founded 2011).

von Ahn was a graduate student at Carnegie Mellon when he was inspired to develop a system for distinguishing between humans and computers.[14] His CAPCTHA solution challenged human users to transcribe distorted letters and prove they weren't computers and spambots. But von Ahn went one step further to turn the CAPTCHA system into a productive use of human computation. In 2006, he was awarded a $500,000 MacArthur genius grant for his proposed idea of using CAPTCHAs to simultaneously prevent scammer bots and to transcribe unreadable printed materials that were indecipherable by optical character recognition (OCR).[15] The newly branded reCAPTCHA successfully completed its first project of digitizing all 13 million articles in *The New York Times'* archive, which starts in 1851.[16] reCAPTCHA was acquired by Google in 2009 and has also been used to contribute to Google Books' goal of digitizing every book in the world (tallying at 25 million books as of 2015).[17] By employing humans to do a task that takes them a mere few seconds but is impossible for computers of today, von Ahn was able to look past his initial technology and find a powerful application for it in society.

---

[11] Ibid.
[12] Ibid.
[13] https://thehustle.co/the-genius-whos-tricking-the-world-into-doing-his-work-recaptcha
[14] Ibid.
[15] Ibid.
[16] Ibid.
[17] https://thehustle.co/the-genius-whos-tricking-the-world-into-doing-his-work-recaptcha

von Ahn's more recent project is Duolingo, a company providing a gamifying platform for users to learn languages for free. Unsurprisingly, von Ahn found yet another creative application of crowdsourcing—three applications, in fact:[18]

1. Users learning languages will practice by translating words or phrases, which are then checked by the software for correctness. If users think the machine graded them incorrectly, they can point out the error and help the model learn. This is the company's solution to the constant evolution of language and ensuring their models remain up to date.

2. Advanced students help translate real pages and articles on the Web. Rather than being graded on textbook translations, they have the opportunity to contribute to translating real material. Duolingo uses principles of crowdsourcing and has multiple translations of the same article to ensure the aggregated translations are accurate. These translations are then used to generate revenue for the company.

3. The Duolingo Incubator provides approved users the opportunity to collaborate and build a language course for others to learn from, thus helping the company generate new educational content.

Through drawing on the motivations and skills of Duolingo's crowd of users, von Ahn has designed an app where his users drive its development simply by using it. This self-powered growth is both smart from a business perspective (generating a revenue stream via application #2 and removing costs of developing course content via application #3) and from a user perspective (providing users the language practice they are seeking via application #1 and encouraging user engagement via application #3).

von Ahn's innovative approaches to human computation have made seemingly impossible tasks possible and remind us all the power and impact that we as a collective can have. While Wikipedia is the default example of crowdsourcing done right, the power of the online crowd can extend beyond just knowledge sharing to algorithm development, consumer product designs, human-powered OCR, and anything else we have the creativity to set our minds to.

---

[18] https://blog.higherlogic.com/duolingo-gets-crowdsourced-community-content-creation

Takeaways: There are numerous applications of crowdsourcing on the Internet today. If companies and individuals can think of creative applications of the collective mind, powerful results such as highly complex algorithmic recommendation engines and transcriptions of previously unreadable printed materials can be achieved.

Questions

1. Consider the impact of the collective group on decision-making (i.e., echo chambers)—what impact do you think there is on individual cognition?

2. Do you frequently use sites such as Wikipedia, Quora, or Stack Exchange? If so, do you trust the information found on these sites? Why or why not?

Dataism and Transhumanism

Religion in the New Age

"God is Dead"—one of the most famous lines in philosophy, declared in 1882 by German philosopher Friedrich Nietzsche, first introduced the idea that society no longer needed a God. Though some interpret this quote as meaning the literal death of God, the phrase was used by Nietzsche to point out that the Age of Enlightenment had displaced the role of religion with science and philosophy. We, as a human society, had outgrown religion.

And yet, the Pew Research Center projected in 2017 that the proportion of religiously unaffiliated people in the world's population would decline in the coming decades.[1] Despite this supposed death of God, it's clear that religion still plays an integral role in our society. As a social-cultural system that prescribes certain morals, beliefs, and behaviors, it is deeply embedded in human culture. Though the exact definition of religion is controversial in religious studies, we can't deny its importance.

The question moving forward in the 21st century is how religion will be newly shaped with the advent of technology. In this chapter, we will look at the various religions that have started relying on technology and then consider the numerous new religions that have emerged in today's digital age.

---

[1] www.pewforum.org/2017/04/05/the-changing-global-religious-landscape/

© Griffin Kao, Jessica Hong, Michael Perusse and Weizhen Sheng 2020
G. Kao et al., *Turning Silicon into Gold*,
https://doi.org/10.1007/978-1-4842-5629-9_25

History of Religion and Technology

The antagonism we thus witness between Religion and Science is the continuation of a struggle that commenced when Christianity began to attain political power.

—John William Draper

The idea that religion and science are incompatible and perpetually in conflict is known as the conflict thesis, first proposed by the Anglo-American scientist and historian John William Draper in the 19th century. The thesis may remain in the public mind when regarding religion and modern science, but it is commonly criticized and discredited by mainstream historians of science for being far too simplistic. In fact, the intertwined histories of religion and science say otherwise, instead suggesting a mutually beneficial and interdependent relation.

Technology is how we take our understanding and knowledge of the sciences and apply them to produce goods or services. There may be similarly perceived discord in the relation between religion and technology and that of religion and science, but once again, a different story emerges when we look at the history of religious institutions and technological developments.

Medieval Europe under Christian rule saw major technological developments, including gunpowder, windmills, mechanical clocks, and more. Irish philosopher Erigena viewed medieval technology, also known as *artes mechanicae* (mechanical arts), to be "man's links to the Divine," proclaiming religious significance on the technology of the day.[2] An illustration of Psalm 63 from the 9th century paints a similar picture, with the godly King David using the first crank, reflecting technological advancements as God's will.[3]

As knowledge progressed toward "perfection" during the Renaissance, a shared mentality that The End was approaching began to emerge, and technology became a part of Christian eschatology (the study of "last things").[4] Science and technology continued to play an important role in religion for many hundreds of years; by the 17th century, many thought that science would lead the way to redemption, and the famed Anglo-Irish chemist Robert Boyle of the time believed that scientists had a special relationship with God.[5] Up until the recent history of the last 200 years or so, religion and the sciences supported and reinforced each other.

---

[2] www.learnreligions.com/technology-as-religion-4038599
[3] https://thefrailestthing.com/2012/02/24/christianity-and-the-history-of-technology-part-three/
[4] www.learnreligions.com/technology-as-religion-4038599
[5] Ibid.

Digitizing Religions Today

As history shows us, technology and religion can certainly coexist harmoniously, and even interdependently. This isn't the common view that we hold today, but looking at the world's religions, it's clear that religions still rely on and interact with technology. Catholic and Protestant churches in the United Kingdom are starting to use contactless card readers for donations, and there exist numerous religious apps such as confession apps (though they don't offer absolution, the act of forgiveness in Christianity for acts of sin), halal apps (to help Muslims find halal restaurants), sermon apps, and more. Moreover, the Vatican has hosted conferences inviting both leaders of faith and leaders in science to discuss the religious and societal impacts of new technologies (i.e., 2018's Unite to Cure: How Science, Technology and 21st Century Medicine Will Impact Culture and Society conference), giving organized religions the opportunity to contribute to the direction of religion and technology's interconnected future narrative.[6]

Some believe that technology will be crucial to the growth of religions. Via the Internet, religions now can now rapidly scale, provide a convenient form of peer surveillance, and promise self-betterment. We have already seen the powerful effects of social media such as Twitter and Reddit for spreading social movements, and these social media platforms will certainly become potentially powerful tools for religions to scale to new audiences as well.

Technology has also made huge amounts of data available at our fingertips, including government social credit systems and records from phone surveillance. This peer surveillance might be reminiscent of that found in religions, offering a form of peer surveillance that can lead to groupthink.[7] This is a phenomenon where a group of people unwittingly arrive at irrational decisions and outcomes due to a desire for conformity—and the vast amounts of today's data make it that much easier to exert religious pressures via shared surveillance and intelligence. A third factor is the idea that technology can lead to the betterment of ourselves, a core aspect of most religions. In the Silicon Valley flavor, people are now hoping to use technology to help humans augment themselves and break past their limits.

The Next Stage of Human Evolution

Transhumanism is the movement pushing for humans to enhance both their minds and bodies using technology. One example of this is genetic engineering, the modification of human genes via biotechnology to improve the human

---

[6] https://techcrunch.com/2018/10/18/disruptive-technology-and-organized-religion/
[7] https://qz.com/1723739/technology-oriented-religions-are-coming/?utm_source=morning_brew

body and treat diseases. Another example is Elon Musk's ambitious Neuralink, a start-up dedicated to linking the human brain directly to a computer by planting a chip in the brain. Although the Pew Research Center has found that highly religious individuals are more against the idea of technologically powered body enhancements,[8] the movement is taking off.

Putting aside ethical concerns, transhumanism offers the opportunity to design ourselves into our ideal image and is often compared to humans attempting to recreate ourselves in God's image. There are limitless possibilities: improved eyesight, higher IQ, and ultimately, the possibility of avoiding death. Silicon Valley is now on a quest for immortality, or "super longevity," with dozens of tech companies such as Calico and Unity Biotechnology attempting to use biotechnology to combat aging and ultimately solve the age-long problem of death.[9]

Another branch of transhumanists seek to escape death by uploading our minds into computers for eternal life via cyberspace. The US entrepreneur Ray Kurzweil believes we will reach a turning point in 2030,[10] when humans and intelligent computers will truly merge, resulting in a powerful and omniscient human–machine mind. The idea that technological growth will one day become uncontrollable and result in computers far beyond human intelligence is often referred to as the singularity.[11]

Although based in science, transhumanism is often compared with religion due to similarities in the pursuit of immortality and spiritual transcendence. In many ways, transhumanism is the modern take on Christian eschatology, where humans themselves are pushing toward the final, "perfect" form of the human mind and body. In fact, transhumanism has religious roots, with the French Jesuit priest Pierre Teilhard de Chardin proposing in 1949 that in the future, all human minds could converge via a global network of machines and reach the divine—finally merging with God.[12]

Today, transhumanism has also fused with traditional religions to yield new ones, intertwining the idea of the singularity with prophecies from religious texts.[13] Dozens of transhumanist churches have arisen, such as the Church of Perpetual Life, the Turing Church, the Christian Transhumanist Association,

---

[8] www.pewresearch.org/fact-tank/2016/07/29/the-religious-divide-on-views-of-technologies-that-would-enhance-human-beings/
[9] www.theguardian.com/technology/2019/feb/22/silicon-valley-immortality-blood-infusion-gene-therapy
[10] https://acton.org/religion-liberty/volume-28-number-4/transhumanism-religion-postmodern-times
[11] www.theguardian.com/technology/2017/apr/18/god-in-the-machine-my-strange-journey-into-transhumanism
[12] www.theguardian.com/technology/2017/apr/18/god-in-the-machine-my-strange-journey-into-transhumanism
[13] Ibid.

and the Mormon Transhumanist Association. These new religious movements build upon traditional religious beliefs as a foundation, with a transhumanist twist to improve ourselves and become closer with God via technology. For example, Mormon Transhumanists seek to "immerse" themselves in the role of Jesus Christ and view expanding upon human abilities via scientific knowledge and technology as the ways to do so.[14] We can clearly see how technology is enhancing existing religious beliefs and providing it new form in the modern world.

The Arrival of a New "God"

But beyond the potential future evolution of humankind, there is the idea that humans won't be the ones to attain the image of God. Instead, as many science fiction authors have imagined throughout the years, the computer will become our new God.

Dataism is a term first used by David Brooks in a 2013 *The New York Times* article and has been discussed by historian Yuval Noah Harari.[15] It refers to the growing dependence of humans on big data, to the point where it has also been compared to religion.[16] The omniscience of data in our societies has prompted the belief that big data will soon know us better than we know ourselves, if not already. As big data, artificial intelligence, and machine learning continue to grow, the idea of the singularity arrives in the picture again.

In 2011, a thought experiment called Roko's Basilisk emerged on an online community.[17] It describes a future, benevolent super-AI which aims to alleviate human suffering. However, by nature of its mission, the AI would punish anyone not supportive of it—including anyone not supporting its existence. Anyone in the past who has heard about it but who then didn't work toward its existence would then be tortured for eternity (apologies for sharing the concept of Roko's Basilisk—you now theoretically need to repurpose your life toward it). As Pascal's wager goes, it would be safer to believe in the existence of God than to be wrong and condemned to Hell. Roko's Basilisk closely resembles traditional religious belief systems, merely replacing the role of God with a super-AI instead.

---

[14] https://transfigurism.org/mission
[15] https://singularityhub.com/2018/09/30/the-rise-of-dataism-a-threat-to-freedom-or-a-scientific-revolution/
[16] Ibid.
[17] www.lastwordonnothing.com/2018/08/10/redux-whos-afraid-of-rokos-basilisk/

In many ways, this super-AI may well be our modern-era God. Although Silicon Valley isn't necessarily proposing the idea of a punishing super-AI as featured in Roko's Basilisk, the tech industry is certainly entertaining the idea of a powerful, omniscient being that is a billion times smarter and more capable than humans. In 2017, Anthony Levandowski, the software engineer known for being caught in a lawsuit between Uber and Waymo, founded a new religion focused on AI called Way of the Future (WOFT).[18] This religion is spreading the idea that technology can and will change our lives and aims to facilitate the transition of humankind to a world reigned by an AI deity.

As Clarke's Third Law (declared by the British science fiction writer Arthur C. Clarke) states, "Any sufficiently advanced technology is indistinguishable from magic." AI is becoming increasingly complex and powerful, and it may soon reach that level of magic beyond the average human's comprehension— that day may mark the true union of religion and technology. Though there is no way of proving whether any of this is science fiction or a future reality, it remains clear that technology is paving the way for a new kind of religion in the modern world.

Takeaways: Technology is a powerful tool for spreading ideas, including religious beliefs. As humans experiment more with technology and augmenting the self, there has been discussion about how far we can push the boundaries of evolution and religion. We must also remain mindful about the potential ramifications of technology in our daily lives and beliefs as big data and AI continue to rapidly grow.

Questions

1.   Has technology ever played a role in shaping your values and beliefs (whether via viral campaigns or online echo chambers)?

2.   Do you think that technology may fundamentally change how we view religion? Will it eventually completely displace current religious systems or simply augment them?

---

[18] www.wired.com/story/anthony-levandowski-artificial-intelligence-religion/

Glossary

Adoption barriers: Obstacles that prevent potential users from using a new product, ranging from minor inconveniences to the need to buy ancillary products to a steep learning curve.

Agent head offices: Banking agents in the M-Pesa outlet network that serve as liaisons between individual outlets and Safaricom; they perform crucial functions like distributing commissions to agents and making sure the agents are well stocked with the cash necessary to service M-Pesa customers.

Ambient computing: Technological environments constantly present and responsive to the needs of users

Angel investor: A person who invests in a new or small business venture, providing capital for start-up or expansion.

Artificial intelligence (AI): The simulation of human intelligence by machines, imitating cognitive functions such as "learning" and "problem-solving"; machine learning (ML) is a key concept associated with AI.

Augmented reality: Technology that superimposes a computer-generated image on a user's view of the real world, thus providing a composite view.

Best-effort deal: When the underwriter for an IPO doesn't buy shares from the company but instead attempts to sell the shares to institutional clients or the investing public.

Big data: Huge volumes of data that cannot be processed with traditional data-processing techniques; an increasingly popular field given that people are producing more and more data on their various devices.

Blue ocean strategy: A marketing theory discussing "value innovation," which is simultaneously pursuing differentiation and low cost to open up new market space.

© Griffin Kao, Jessica Hong, Michael Perusse and Weizhen Sheng 2020
G. Kao et al., *Turning Silicon into Gold*,
https://doi.org/10.1007/978-1-4842-5629-9

Bought deal: When the underwriter for an IPO buys shares from the company and sells the shares to the public, thereby taking on the risk of unsold shares.

Brick and mortar: A term used to refer to stores with a physical, customer-facing location as opposed to an online presence.

Carbon neutral: Making no net release of carbon dioxide to the atmosphere, especially through offsetting emissions by planting trees.

Churn: The percentage of users that stop using a product or service during a certain period of time.

Click-through rate (CTR): The rate at which users who see a visual element like an ad or button will click on it—a commonly used metric to gauge how successful particular features are.

Conglomerate: A combination of multiple business entities operating in entirely different industries under one corporate group, usually involving a parent company and many subsidiaries

Continuation funds: A fund to which general partners will move assets to from an older fund in a fund restructuring deal. They can provide general partners acting as portfolio managers greater control over the portfolio and the ability to maximize the potentially underperforming remaining assets.

Conversion rate: The number of conversions divided by the total number of users.

Critical mass: In regard to the network effect, the amount of users needed on platform for the value of the product or service to equal or outweigh the cost of the product.

Crowdsourcing: The sourcing of work and/or ideas from Internet users;a portmanteau of the words "crowd" and "outsourcing"

Customer acquisition: The process of obtaining new customers, usually a focus for emerging and growing companies.

Customer conversion funnel: The process of converting people into paying and loyal customers, where each subsequent stage of the process captures a smaller subset (see Figure 3-2).

Customer lifetime value: The predicted net profit from the entire future relationship with the customer.

Customer segment: A group of individuals that are similar in specific ways such as age, gender, interests, and spending habits, which are relevant to marketing.

Cyberspace: A concept of a widespread, interconnected technology environment; a term that represents the many new technological ideas and phenomena that are emerging.

D2C: the acronym for "direct to consumer"; referring to companies (commonly e-commerce ones) selling directly to their consumers.

Dataism: The growing dependence of humans on big data; often compared to religion.

Deep learning: A family of machine learning methods using artificial neural networks.

Defensible: A product that is defensible is one that is hard to replace by any other competing products.

Demand-side platform (DSP): Software that enables one to manage an advertising campaign across many real-time bidding networks like Google or Facebook Ads.

Direct (ownership) stake: Equity held in a company as a percentage. Founders and initial investors usually hold direct ownership stakes.

Direct secondary fund: A type of secondary fund that primarily performs direct secondary transactions.

Direct secondary: A type of secondary deal in which the holder of the shares in a private company, such as a founder or employee or investor, sells their directly held equity to some buyer.

Distribution virality: Product virality achieved by users spreading awareness of the product within their existing network by sharing the outcome of the product.

Emerging market: Developing countries with high economic growth prospects that are becoming more and more engaged with global markets.

Enterprise software: Computer software that is used to satisfy the needs of an organization or business rather than individual users.

Equity dilution: The decrease in existing shareholders' ownership of a company as a result of the company issuing new equity.

Equity interest stake: Generally used to refer to any kind of holding in a private company—this can be a direct ownership stake or a stake in a venture capital fund that then maintains shares in portfolio companies (a fund interest).

Follow-on investments: Investments made by an investor who made a previous investment in the company, which are often required as a result of "pay-to-play" provisions in the term sheets governing venture capital deals.

Freemium: A common pricing strategy where a product is offered free of charge, but additional features, upgrades, or expansions are offered for an additional price.

Friction: A barrier to product adoption, growth, or usage.

Fund restructuring: A type of secondary deal in which a general partner typically moves assets or shares from funds exceeding their expected life cycles to new vehicles (i.e., a new fund with different terms). Limited partners participating in the older fund can choose to cash out and sell their stake or transfer their shares to the new special-purpose vehicle.

Gen Z: Also known as Generation Z; the demographic cohort following the Millennials, broadly defined as people born from the mid- to late 1990s until 2010

General partner: A partner that can be seen as the owner of the partnership, actively managing the venture's day-to-day operations. In exchange for this power, they hold unlimited financial liability in the case of some adverse event like a lawsuit.

Genetic engineering: Directly manipulating an organism's genes using biotechnology.

Graphical user interface (GUI): A user interface allowing users to interact with technology via a graphical screen with graphical elements (i.e., icons).

Head-mounted display: A display device worn on the head.

Human computation: A computer science technique that outsources certain steps to humans.

Human–computer interaction: The research of the design and use of computers and technology, with a focus on how it interfaces with users

Hybrid secondary fund: A type of secondary fund that acquires both indirect interests and performs direct secondary transactions.

Initial public offering (IPO): The process in which a company offers shares of a private corporation to the public in a new stock issuance, thereby raising capital for the firm to use.

Innovation: The application of better solutions in products and/or services which add value via new ideas or creative thoughts.

Last mile delivery problem: The issue online retailers and shipping services have to contend with where the last leg of the delivery is disproportionately costly when compared with the rest of the journey.

Leverage: The use of debt or borrowed capital to undertake an investment or project.

Limited partner fund: A type of secondary fund that primarily acquires indirect partnership interests through venture funds.

Limited partner: A partner who is removed from the daily proceedings of the company but maintains some stake in the venture. In general, limited partners are only liable for the amount of capital they invest in the business but are privy to some proportional percentage of the partnership's profits.

Liquidity: The ease at which a security can be turned into liquid assets or cash.

Long-term viability: A company's long-term survival and its ability to sustain profits over a period of time.

Machine learning (ML) model: A mathematical model that computationally makes predictions by finding patterns in data.

Market cannibalization: A loss in sales caused by a company's introduction of a new product that displaces one of its own older products.

Market penetration: The successful selling of a product or service in a specific market as measured by the amount of sales volume of the product or service in relation to the total target market size.

Market saturation: When the volume of a product or service sold in a market has been maximized; this often happens when demand of the product decreases.

Market share: The portion of a market controlled by a particular company or product.

Marketplace (two-sided platform): A type of e-commerce site that connects potential buyers and sellers all within one platform to help rent, buy, swap, or negotiate.

Massively multiplayer online game: An online game with large numbers of players who can participate simultaneously over the Internet.

Meme: A unit of culture, such as an idea, belief pattern, or behavior, that spreads from person to person.

Microfinancing: A term used to describe financial services, such as loans, savings, insurance, and fund transfers, to entrepreneurs, small businesses, and individuals who lack access to traditional banking services.

Monetization: The process of turning a product or service into a source of revenue.

M-Pesa: Originating in Kenya, one of the first popularized mobile payment systems.

Multiple expansion: A strategy in which an investor gets in on the venture and can flip their stake for some multiple of their initial investment in the absence of true appreciation in value because some form of superficial growth in the company warrants a greater perceived value.

Multisided platform: A model in which two or more separate parties come together to transact on some network.

Natural language processing (NLP): A field intersecting linguistics, computer science, machine learning, and artificial intelligence that studies the interactions between computers and human (natural) languages.

Network effect: The phenomenon whereby increased numbers of people or participants increases the value of a good or service for others.

Newtonian engagement: A concept stating that an engaged player of freemium service will remain engaged until acted upon by an outside force.

Next billion users: A classification of the segment of tech users getting online for the first time across many developing countries.

O2O: The acronym for "online to offline"; referring to the mix of both online digital environments and physical locations in retail.

Offering price: The share price during an IPO that the underwriter sets, usually only available to institutional clients and dependent on the perceived demand/capital the company would like to raise.

Opening price: The share price of an IPOing company on the first day the company's shares are available to the public, based on supply and demand.

Original content: Creative material (such as a TV show) that is unique, new, and owned by the individual creating it.

Partnership: When multiple parties come together to manage or own a venture—this entity requires no filing fees nor registration requirements but has important implications for how liability is held.

Personal digital assistant: A mobile device providing information manager functions; now mostly displaced by smartphones.

Personalization: Tailoring a service/product to the user; a trend in technology and social media.

Pivot: A change in a company's strategy or direction.

Planned obsolescence: A policy of producing consumer goods that rapidly become obsolete and so require replacing, achieved by frequent changes in design, termination of the supply of spare parts, and the use of nondurable materials.

Preferred equity: A type of secondary deal in which a general partner provides a debt-like interest as preferred equity to buyers.

Primary market: The market where assets or real goods are created and first issued. In the context of venture capital, this is when equity is first obtained by a venture capital firm in exchange for funding.

Product life cycle: The four stages consisting of introduction, growth, maturity, and decline that a product experiences during its life on the market.

Profitless prosperity model: The business model increasingly employed by companies in the growth phase, in which the company focuses less on generating a profit and more on quickly increasing market share (in the hopes of later being able to reduce cost or convert users to paid ones/higher-paying ones).

Pull virality: Product virality achieved by users who invite people from their network to use the product in order to gain more value from the product.

Qualitative user research: Research done through observation and non-numerical insights (such as personal opinions or experiences) to better understand user behavior and needs.

Quantitative metrics: Numerical data points businesses use to track, monitor, and assess the success or failure of various business processes.

Recommendation system: A subclass of information-filtering systems that seeks to predict the preference a user would give to an item.

Research and development (R&D): Research, sometimes resource-intensive, undertaken by firms in developing new services or products or improving existing services or products.

Royalties: A sum of money paid to a patentee for the use of a patent or other work.

Scalable: Able to perform well under an increased or expanding scope.

Second mover advantage: The advantage a company obtains by following others into a market.

Secondary market: The market consisting of the subsequent exchange of commodities between parties independent of the initial issuer. In the context of venture capital, the secondary market consists of various transactions between investors involving already issued interests in a private venture.

Security: An asset pledged as a guarantee of the fulfillment of an undertaking or the repayment of a loan—often used to refer to bonds or stocks.

Singularity: The idea that technological growth will one day become uncontrollable and result in computers far beyond human intelligence.

SMS: The acronym for "Short Message Service"; a text messaging service using standardized communication protocols that is currently being challenged by the rise of social messaging apps.

Social messaging app: Internet protocol-based apps which support broad platforms for messaging; widely used on smartphones.

Speech recognition: A field in computational linguistics that enables computers to recognize and translate spoken language using knowledge from linguistics, computer science, machine learning, and electrical engineering

Stake: Some percentage ownership in a company or venture that entitles the holder to receive a portion of the profits/growth commensurate to the size of their stake. Venture capital firms usually ask for some stake in the venture they're funding.

Stapled secondary: A type of secondary deal that is usually just fund restructuring or tender offer deals where the general partner will cleverly add another requirement for the purchasing parties to also commit fresh, additional capital to a new fund.

Sticky: A sticky product is a product that is highly memorable and generates repeat usage.

Stock split: An issue of new shares in a company to existing shareholders in proportion to their current holding.

Streaming rights: The official legal ability to play video content on a certain platform.

Subscription model: A business model in which a customer must pay a recurring price at regular intervals for access to a product or service.

Suicide algorithms: The informal name for Facebook's "proactive suicide detection algorithms" that assess user data to predict mental health risks.

Super app: An umbrella app containing multiple apps that can be compared to an operating system for life due to the numerous functions offered (i.e., messaging, payments, food delivery); common in Asian cities with examples such as WeChat, Kakao, and LINE.

Super-voting stock class: A special type of company stock that holds above-average voting rights associated with it, typically owned by company founders and early backers and not traded publicly.

Supply-side platform (SSP): A network-specific tool, like Facebook Connect, that enables the application of complex algorithm to automate the process of purchasing and posting successful targeted ads to social media platforms and new sites at scale (programmatic advertising).

Targeted content: Specific content created and presented for a subset of a wider audience.

Techlash: Portmanteau of "tech" and "backlash;" a strong reaction against the major technology companies, as a result of concerns about their power, users' privacy, the possibility of political manipulation, and other controversial matters.

Tender offer: A type of secondary deal in which a general partner holds an auction (called "tendering") for currently held interests to incur offers from secondary buyers. Limited partners holding those interests can then choose to liquidate or maintain their stake in the fund.

Third-party market: The secondhand sale of a good outside of its initial producer's markets.

Tragedy of the commons: The depletion of a resource because a large number of individuals act independently, according to self-interest, and contrary to the common good.

Transhumanism: A movement pushing for humans to enhance and augment both their minds and bodies using technology.

Underwriter: Any party that evaluates and assumes another party's risk for a fee, such as the investment bank that buys a company's shares ahead of an IPO.

User acquisition: The process of obtaining new users.

User-generated content: Any form of content, such as text and images, that are produced by users of an online platform (i.e., reviews, ratings).

User retention: The ability to keep users over a period of time.

Venture capital: A type of private equity, a form of financing that is provided by firms or funds to small, early-stage, and/or emerging firms that are considered to have demonstrated or have the potential for high growth.

Venture exits: A public stock exchange or get acquired by another company for a large sum of money.

Vertical integration: A strategy in which a company attempts to own multiple parts of the supply chain, each contributing to a different stage of the product, in an effort to reduce cost and increase control.

Viral coefficient: The number of new users generated by an existing user.

Viral marketing: A method of marketing whereby consumers are encouraged to share information about a company's goods or services via the Internet.

Virality: The tendency of consumers to spread information about products through their social networks.

Virtual assistant: Software that performs tasks for individuals based on commands (typically voice commands). Also commonly referred to as voice assistant, smart assistants, or intelligent personal assistants.

Virtual reality: A computer-generated simulation of an environment that can be interacted with.

Voice user interface (VUI): a user interface allowing users to interact with technology via voice commands, using speech recognition technology.

Voting shares: A special type of company share that gives the stockholder the right to vote on matters of corporate policy-making.

Wisdom of the crowd: the phenomenon where the collective thoughts and knowledge of a group of individuals is on par, if not better, than that of a single expert.

Y Combinator: A prestigious American seed start-up accelerator.

Index

© Griffin Kao, Jessica Hong, Michael Perusse and Weizhen Sheng 2020
G. Kao et al., *Turning Silicon into Gold*,
https://doi.org/10.1007/978-1-4842-5629-9

W, X

Y, Z

CPSIA information can be obtained
at www.ICGtesting.com
Printed in the USA
BVHW040208090320
574503BV00004B/93